Energy Conservation Handbook

Energy Conservation Handbook

Samuel Palmer

R CALLISTO REFERENCE

www.callistoreference.com

Callisto Reference,
118-35 Queens Blvd., Suite 400,
Forest Hills, NY 11375, USA

Visit us on the World Wide Web at:
www.callistoreference.com

ISBN: 978-1-64116-595-2 (Hardback)

Cataloging-in-Publication Data

Energy conservation handbook / Samuel Palmer.
  p. cm.
Includes bibliographical references and index.
ISBN 978-1-64116-595-2
1. Energy conservation. 2. Energy consumption. 3. Energy policy.
4. Conservation of natural resources. I. Palmer, Samuel.
QC73.8.C6 E54 2022
333.791 6--dc23

TABLE OF CONTENTS

Permissions

Index

PREFACE

The purpose of this book is to help students understand the fundamental concepts of this discipline. It is designed to motivate students to learn and prosper. I am grateful for the support of my colleagues. I would also like to acknowledge the encouragement of my family.

Energy conservation refers to the reduction in the consumption of energy through the efficient use of energy as well as the reduction in the amount of services used. Energy conservation seeks to increase the environmental quality. It also reduces energy costs by preventing future resource depletion. One of the major ways of improving conservation of energy in buildings is through conducting an energy audit. It is concerned with an analysis and inspection of energy flow and consumption for energy conservation in a building, system or process. This book unfolds the innovative aspects of energy conservation which will be crucial for the holistic understanding of the subject matter. Most of the topics introduced herein cover new techniques of energy conservation. This book aims to serve as a resource guide for students and experts alike and contribute to the growth of the discipline.

A foreword for all the chapters is provided below:

Chapter – Introduction

Energy is the quantitative property that is transferred to an object for heating it or performing work on it. Some of the major areas of study related to energy are energy conservation, energy transformation and energy transition. This chapter will briefly introduce all these significant aspects related to energy.

Chapter – Sources of Energy

Sources of energy are broadly classified into renewable sources and non-renewable sources. The renewable resources of energy include bioenergy, hydropower and geothermal energy while non-renewable sources of energy are fossil fuels such as coal, natural gas and petroleum. The topics elaborated in this chapter will help in gaining a better perspective about these sources of energy.

Chapter – Energy Consumption and Efficiency

The total amount of energy or power used is termed as energy consumption. Energy efficiency refers to the practices which are aimed at reducing the wastage of energy required to provide products and services. The chapter closely examines the key concepts of energy consumption and efficiency such as energy monitoring and energy crisis to provide an extensive understanding of the subject.

Chapter – Strategies for Energy Conservation

The major strategies of energy conservation include energy harvesting, energy recovery, green building, zero energy building, energy development, green computing, and using smart grids, alternate energy and passive houses. These strategies of energy conservation have been thoroughly discussed in this chapter.

Chapter – Sustainable Energy

Sustainable energy is a practice that focuses on fulfilling the present demands of energy without jeopardizing the ability of future generation to meet their needs. The efforts which are made in order to reduce the consumption of energy by lessening the usage of a particular energy service are termed as energy conservation. The diverse aspects of sustainable energy and energy conservation have been thoroughly discussed in this chapter.

Samuel Palmer

Introduction   1

Energy is the quantitative property that is transferred to an object for heating it or performing work on it. Some of the major areas of study related to energy are energy conservation, energy transformation and energy transition. This chapter will briefly introduce all these significant aspects related to energy.

Energy

Energy is the capacity for doing work. It may exist in potential, kinetic, thermal, electrical, chemical, nuclear, or other various forms. There are, moreover, heat and work—i.e., energy in the process of transfer from one body to another. After it has been transferred, energy is always designated according to its nature. Hence, heat transferred may become thermal energy, while work done may manifest itself in the form of mechanical energy.

All forms of energy are associated with motion. For example, any given body has kinetic energy if it is in motion. A tensioned device such as a bow or spring, though at rest, has the potential for creating motion; it contains potential energy because of its configuration. Similarly, nuclear energy is potential energy because it results from the configuration of subatomic particles in the nucleus of an atom.

Energy can be neither created nor destroyed but only changed from one form to another. This principle is known as the conservation of energy or the first law of thermodynamics. For example, when a box slides down a hill, the potential energy that the box has from being located high up on the slope is converted to kinetic energy, energy of motion. As the box slows to a stop through friction, the kinetic energy from the box's motion is converted to thermal energy that heats the box and the slope.

Energy can be converted from one form to another in various other ways. Usable mechanical or electrical energy is, for instance, produced by many kinds of devices, including fuel-burning heat engines, generators, batteries, fuel cells, and magnetohydrodynamic systems.

In the International System of Units (SI), energy is measured in joules. One joule is equal to the work done by a one-newton force acting over a one-metre distance.

Conservation of Energy

Conservation of energy is the principle according to which the energy of interacting bodies or particles in a closed system remains constant. The first kind of energy to be recognized was kinetic energy, or energy of motion. In certain particle collisions, called elastic, the sum of the kinetic energy of the particles before collision is equal to the sum of the kinetic energy of the particles after collision. The notion of energy was progressively widened to include other forms. The kinetic energy lost by a body slowing down as it travels upward against the force of gravity was regarded as being converted into potential energy, or stored energy, which in turn is converted back into kinetic energy as the body speeds up during its return to Earth. For example, when a pendulum swings upward, kinetic energy is converted to potential energy. When the pendulum stops briefly at the top of its swing, the kinetic energy is zero, and all the energy of the system is in potential energy. When the pendulum swings back down, the potential energy is converted back into kinetic energy. At all times, the sum of potential and kinetic energy is constant. Friction, however, slows down the most carefully constructed mechanisms, thereby dissipating their energy gradually. During the 1840s it was conclusively shown that the notion of energy could be extended to include the heat that friction generates. The truly conserved quantity is the sum of kinetic, potential, and thermal energy. For example, when a block slides down a slope, potential energy is converted into kinetic energy. When friction slows the block to a stop, the kinetic energy is converted into thermal energy. Energy is not created or destroyed but merely changes forms, going from potential to kinetic to thermal energy. This version of the conservation-of-energy principle, expressed in its most general form, is the first law of thermodynamics. The conception of energy continued to expand to include energy of an electric current, energy stored in an electric or a magnetic field, and energy in fuels and other chemicals. For example, a car moves when the chemical energy in its gasoline is converted into kinetic energy of motion.

With the advent of relativity physics, mass was first recognized as equivalent to energy. The total energy of a system of high-speed particles includes not only their rest mass but also the very significant increase in their mass as a consequence of their high speed. After the discovery of relativity, the energy-conservation principle has alternatively been named the conservation of mass-energy or the conservation of total energy.

When the principle seemed to fail, as it did when applied to the type of radioactivity called beta decay (spontaneous electron ejection from atomic nuclei), physicists accepted the existence of a new subatomic particle, the neutrino, that was supposed to carry off the missing energy rather than reject the conservation principle. Later, the neutrino was experimentally detected.

Energy conservation, however, is more than a general rule that persists in its validity. It can be shown to follow mathematically from the uniformity of time. If one moment of time were peculiarly different from any other moment, identical physical phenomena occurring at different moments would require different amounts of energy, so that energy would not be conserved.

Energy Transformation

Energy transformation, also known as energy conversion, is the process of changing energy from one form to another. In physics, energy is a quantity that provides the capacity to perform work (e.g. lifting an object) or provides heat. In addition to being convertible, according to the law of conservation of energy, energy is transferable to a different location or object, but it cannot be created or destroyed.

Energy in many of its forms may be used in natural processes, or to provide some service to society such as heating, refrigeration, lighting or performing mechanical work to operate machines. For example, in order to heat a home, the furnace burns fuel, whose chemical potential energy is converted into thermal energy, which is then transferred to the home's air to raise its temperature.

Limitations in the Conversion of Thermal Energy

Conversions to thermal energy from other forms of energy may occur with 100% efficiency . Conversion among non-thermal forms of energy may occur with fairly high efficiency, though there is always some energy dissipated thermally due to friction and similar processes. Sometimes the efficiency is close to 100%, such as when potential energy is converted to kinetic energy as an object falls in a vacuum. This also applies to the opposite case; for example, an object in an elliptical orbit around another body converts its kinetic energy (speed) into gravitational potential energy (distance from the other object) as it moves away from its parent body. When it reaches the furthest point, it will reverse the process, accelerating and converting potential energy into kinetic. Since space is a near-vacuum, this process has close to 100% efficiency.

Thermal energy is unique because it can not be converted to other forms of energy. Only a difference in the density of thermal energy (temperature) can be used to perform work, and the efficiency of this conversion will be (much) less than 100%. This is because thermal energy represents a particularly disordered form of energy; it is spread out randomly among many available states of a collection of microscopic particles constituting the system (these combinations of position and momentum for each of the particles are said to form a phase space). The measure of this disorder or randomness is entropy, and its defining feature is that the entropy of an isolated system never decreases. One cannot take a high-entropy system (like a hot substance, with a certain amount of thermal energy) and convert it into a low entropy state (like a low-temperature substance, with correspondingly lower energy), without that entropy going somewhere else (like the surrounding air). In other words, there is no way to concentrate energy without spreading out energy somewhere else.

Thermal energy in equilibrium at a given temperature already represents the maximal evening-out of energy between all possible states. Such energy is sometimes called "degraded energy," because it is not entirely convertible a "useful" form, i.e. one that can do more than just affect temperature. The second law of thermodynamics states that the entropy of a closed system can never decrease. For this reason, thermal energy in a system may be converted to other kinds of energy with efficiencies approaching 100% only if the entropy of the universe is increased by other means, in order to compensate for the decrease in entropy associated with the disappearance of the thermal energy and its entropy content. Otherwise, only a part of that thermal energy may be converted to other kinds of energy (and thus useful work). This is because the remainder of the heat must be reserved to be transferred to a thermal reservoir at a lower temperature. The increase in entropy for this process is greater than the decrease in entropy associated with the transformation of the rest of the heat into other types of energy.

In order to make energy transformation more efficient, it is desirable to avoid thermal conversion. For example, the efficiency of nuclear reactors, where the kinetic energy of the nuclei is first converted to thermal energy and then to electrical energy, lies at around 35%. By direct conversion of kinetic energy to electric energy, effected by eliminating the intermediate thermal energy transformation, the efficiency of the energy transformation process can be dramatically improved.

Examples of Sets of Energy Conversions in Machines

A coal-fired power plant involves these energy transformations:

1. Chemical energy in the coal converted to thermal energy in the exhaust gases of combustion.

2. Thermal energy of the exhaust gases converted into thermal energy of steam through heat exchange.

3. Thermal energy of steam converted to mechanical energy in the turbine.

4. Mechanical energy of the turbine converted to electrical energy by the generator, which is the ultimate output.

In such a system, the first and fourth steps are highly efficient, but the second and third steps are less efficient. The most efficient gas-fired electrical power stations can achieve 50% conversion efficiency. Oil- and coal-fired stations are less efficient.

In a conventional automobile, the following energy transformations occur:

1. Chemical energy in the fuel converted to kinetic energy of expanding gas via combustion.

2. Kinetic energy of expanding gas converted to linear piston movement.

3. Linear piston movement converted to rotary crankshaft movement.

4. Rotary crankshaft movement passed into transmission assembly.

5. Rotary movement passed out of transmission assembly.

6. Rotary movement passed through differential.

7. Rotary movement passed out of differential to drive wheels.

8. Rotary movement of drive wheels converted to linear motion of the vehicle.

Other Energy Conversions

Lamatalaventosa Wind Farm.

There are many different machines and transducers that convert one energy form into another. A short list of examples follows:

- Thermoelectric (Heat → Electrical energy).

- Geothermal power (Heat→ Electrical energy).

- Heat engines, such as the internal combustion engine used in cars, or the steam engine (Heat → Mechanical energy).

- Ocean thermal power (Heat → Electrical energy).

- Hydroelectric dams (Gravitational potential energy → Electrical energy).

- Electric generator (Kinetic energy or Mechanical work → Electrical energy).

- Fuel cells (Chemical energy → Electrical energy).

- Battery (electricity) (Chemical energy → Electrical energy).

- Fire (Chemical energy → Heat and Light).

- Electric lamp (Electrical energy → Heat and Light).

- Microphone (Sound → Electrical energy).

- Wave power (Mechanical energy → Electrical energy).

- Windmills (Wind energy → Electrical energy or Mechanical energy).

- Piezoelectrics (Strain → Electrical energy).

- Friction (Kinetic energy → Heat).

- Electric heater (Electric energy → Heat).

- Photosynthesis (Electromagnetic radiation → Chemical energy).

- ATP hydrolysis (Chemical energy in adenosine triphosphate → mechanical energy).

Energetics

Energetics (also called energy economics) is the study of energy under transformation. Because energy flows at all scales, from the quantum level to the biosphere and cosmos, energetics is a very broad discipline, encompassing for example thermodynamics, chemistry, biological energetics, biochemistry and ecological energetics. Where each branch of energetics begins and ends is a topic of constant debate. For example, Lehninger contended that when the science of thermodynamics deals with energy exchanges of all types, it can be called energetics.

In general, energetics is concerned with defining relationships to describe the tendencies of energy flows and storages under transformation, defined here as phenomena which behave like historical invariants under repeated observations. When some critical number of people have observed such invariance, such a principle is usually then given the status of a 'fundamental law' of science. As in all scientific inquiry, whether a theorem or principle is considered a fundamental law appears to depend on how many people agree to the proposition. The ultimate aim of energetics therefore is the description of fundamental laws. Philosophers of science have held that the fundamental laws of thermodynamics can be treated as laws of energetics, by continuing to more accurately describe these laws, energetics aims to produce reliable predictions about energy flow and storage transformations at any scale.

Energetics has a controversial history. Some authors maintain that the its origins may be found in the work of the ancient Greeks, but that the mathematical formalisation began with the work of Leibniz. Richard de Villamil said that Rankine formulated the science of energetics. W. Ostwald and E. Mach subsequently developed the study, and by the late 1800s energetics was understood to be incompatible with the atomic view of the atom forwarded by Boltzmann's gas theory. Proof of the atom settled the dispute but not without significant damage. In the 1920s Lotka attempted to build on Boltzmann's views through a mathematical synthesis of energetics with biological evolutionary theory. Lotka proposed that the selective principle of evolution was one which favoured the maximum useful energy flow transformation. This view subsequently influenced the further development of ecological energetics, especially the work of Howard T. Odum.

De Villamil attempted to clarify the scope of energetics with respect to other branches of physics by positing a system that divides mechanics into two branches; energetics (the science of energy), and "pure", "abstract" or "rigid" dynamics (the science of momentum). According to Villamil energetics can be mathematically characterised by scalar equations, and rigid dynamics by vector equations. In this division the dimensions for dynamics are *space*, time and mass, and for energetics, *length*, time and mass. This division is made according to fundamental suppositions about the properties of bodies, e.g.:

1. Are the particles comprising the system rigidly fixed together?

2. Is there any machinery for stopping moving bodies?

In Villamil's classification system, dynamics says yes to 1 and no to 2, whereas energetics says no to 1 and yes to 2. Therefore, in Villamil's system, dynamics assumes that particles are rigidly fixed together and cannot vibrate, and consequently must all be at zero kelvin. The conservation of momentum is a consequence of this view, however it is considered valid only in logic and not to be a true representation of the facts. In contrast energetics does not assume that particles are rigidly fixed together, and thus are free to vibrate, and consequently can be at non-zero temperatures.

Principles of Energetics

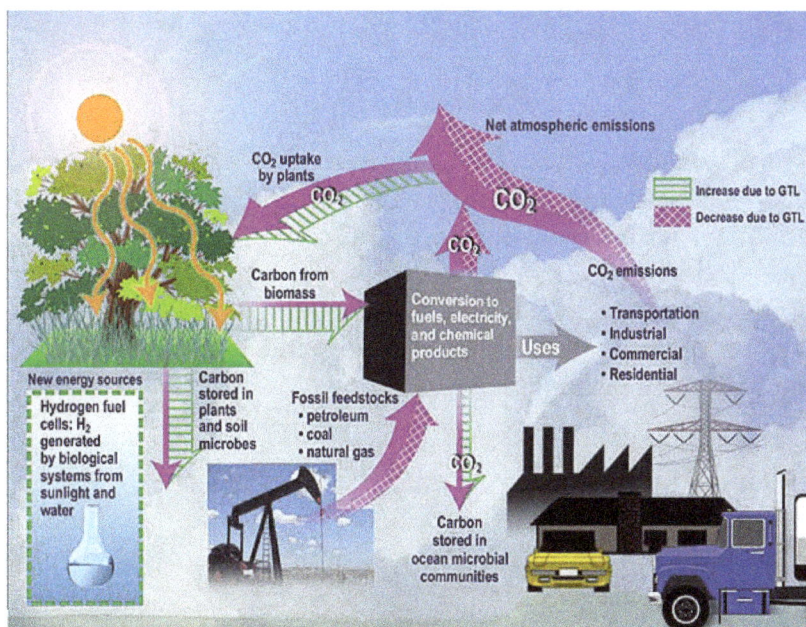

Ecological analysis of CO_2 in an ecosystem.

As a general statement of energy flows under transformation, the principles of energetics include the first four laws of thermodynamics which seek a rigorous description. However the precise place of the laws of thermodynamics within the principles of energetics is a topic currently under debate. If the ecologist Howard T. Odum was right, the principles of energetics take into consideration a hierarchical ordering of energy forms, which aims to account for the concept of energy quality, and the evolution of the universe. Albert Lehninger called these hierarchical orderings the "successive stages in the flow of energy through the biological macrocosm".

Odum proposed 3 further energetic principles and one corollary that take energy hierarchy into account. The first four principles of energetics are related to the same numbered laws of thermodynamics.

Energy Transition

Energy transition is generally defined as a long-term structural change in energy systems. These have occurred in the past, and still occur worldwide. Historic energy transitions are most broadly described by Vaclav Smil. Contemporary energy transitions differ in terms of motivation and objectives, drivers and governance.

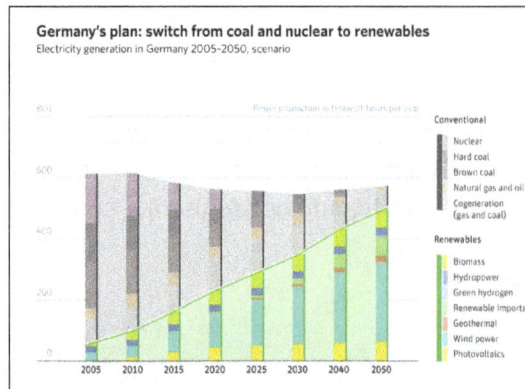

Energy transition scenario.

The layout of the world's energy systems has changed significantly over time. Until the 1950s, the economic mechanism behind energy systems was local rather than global. As development progressed, different national systems became more and more integrated becoming the large, international systems seen today. Historical transition rates of energy systems have been extensively studied. While historical energy transitions were generally protracted affairs, unfolding over many decades, this does not necessarily hold true for the present energy transition, which is unfolding under very different policy and technological conditions.

Photovoltaic array and wind turbines at the Schneebergerhof
wind farm in the German state of Rheinland-Pfalz.

Solving the global warming problem is regarded as the most important challenge facing humankind in the 21st century. The capacity of the earth system to absorb greenhouse gas emissions is already exhausted, and under the Paris climate agreement, emissions must cease by 2040 or 2050. Barring a breakthrough in carbon sequestration technologies, this requires an energy transition away from fossil fuels such as oil, natural gas, lignite, and coal. This energy transition is also known as the decarbonization of the energy system. Available technologies are nuclear power (fission) and the renewable energy sources wind, hydropower, solar power, geothermal, and marine energy.

A timely implementation of the energy transition requires multiple approaches in parallel. Energy conservation and improvements in energy efficiency thus play a major role. Smart electric meters can schedule energy consumption for times when electricity is abundant, reducing consumption at times when the more variable renewable energy sources are scarce (night time and lack of wind).

Parabolic trough power station for electricity production, near the town of
Kramer Junction in California's San Joaquin Valley.

Despite the widespread understanding that a transition to renewable energy is necessary, there are a number of risks and barriers to making renewable energy more appealing than conventional energy. Renewable energy rarely comes up as a solution beyond combating climate change, but has wider implications for food security and employment. This further supports the recognized dearth of research for clean energy innovations, which may lead to quicker transitions. Overall, the transition to renewable energy requires a shift among governments, business, and the public. Altering public bias may mitigate the risk of subsequent administrations de-transitioning - through perhaps public awareness campaigns or carbon levies.

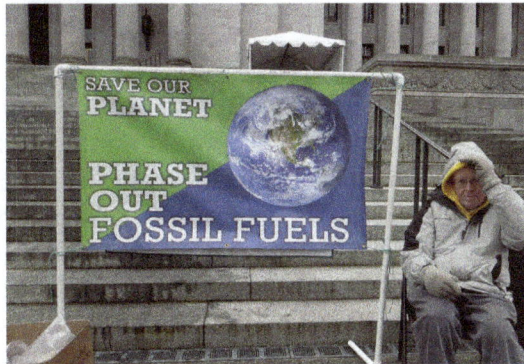

Protest at the Legislative Building in Olympia, Washington. Ted Nation, activist for several decades, beside sign reading "Save our planet. Phase out Fossil Fuels".

After a transitional period, renewable energy production is expected to make up most of the world's energy production. The risk management firm, DNV GL, forecasts that, by 2050, the world's primary energy mix will be split equally between fossil and non-fossil sources.

Energy Transition Timeline

A 2011 projection expects solar PV to supply more than half of the world's electricity by 2060, dramatically reducing the emissions of greenhouse gases.

An example of transition toward sustainable energy, is the shift by Germany and Switzerland, to decentralised renewable energy, and energy efficiency. Although so far these shifts have been replacing nuclear energy, their declared goal 2012 was the abolishment of coal, reducing non-renewable energy sources and the creation of an energy system based on 60% renewable energy by 2050. As of 2018, the 2030 coalition goals are to achieve 65% renewables in electricity production until 2030 in Germany.

An 'energy transition' designates a significant change for an energy system that could be related to one or a combination of system structure, scale, economics, and energy policy. An 'energy transition' is usefully defined as a change in the state of an energy system as opposed to a change in an individual energy technology or fuel source. A prime example is the change from a pre-industrial system relying on traditional biomass and other renewable power sources (wind, water, and muscle power) to an industrial system characterized by pervasive mechanization (steam power) and the use of coal. Market shares reaching pre-specified thresholds are typically used to characterize the speed of transition (e.g. coal versus traditional biomass) and typical market share thresholds in the literature are 1%, 10% for the initial shares and 50%, 90% and 99% for outcome shares following a transition.

For energy systems, many lessons can be learned from history. The need for large amounts of firewood in early industrial processes in combination with prohibitive costs for overland transportation led to a scarcity of accessible (e.g. affordable) wood and it has been found that eighteenth century glass-works "operated like a forest clearing enterprise. When Britain had to resort to coal after largely having run out of wood, the resulting fuel crisis triggered a chain of events that two centuries later culminated in the Industrial Revolution. Similarly, increased use of peat and coal was vital elements paving the way for the Dutch Golden Age roughly spanning the entire 17th century. Another example where resource depletion triggered technological innovation and a shift to new energy sources in 19th Century whaling and how whale oil eventually became replaced by kerosene and other petroleum-derived products.

Technology has been identified as an important but difficult-to-predict driver of change within energy systems. Published forecasts have systematically tended to overestimate the potential of new energy and conversion technologies and underestimated the inertia in energy systems and energy infrastructure (e.g. power plants, once built, characteristically operate for many decades). The history of large technical systems is very useful for enriching debates about energy infrastructures by detailing many of their long-term implications. The speed at which a transition in the energy sector needs to take place will be historically rapid. Moreover, the underlying technological, political and economic structures will need to change radically — a process one author calls regime shift.

The term 'energy transition' could also encompasses a reorientation of policy and this is often the case in public debate about energy policy. For example, this could imply a rebalance of demand to supply and a shift from centralized to distributed generation (for example, producing heat and power in very small cogeneration units), which should replace overproduction and avoidable energy consumption with energy-saving measures and increased efficiency. In a broader sense the energy transition could also entail a democratization of energy or a move towards increased sustainability.

In June 2018, at their G20 Summit in Argentina, the G20 Energy Ministers 'welcome(d) the approach of Argentina's G20 Presidency, which recognises that there are different possible national paths to achieve cleaner energy systems - while promoting sustainability, resilience and energy security - under the term "transitions" (in plural). This view reflects the fact that each G20 member - according to its stage of development - has a unique and diverse energy system as starting point, with different energy resources, demand dynamics, technologies, stock of capital, geographies and cultures.'

Status in Specific Countries

Australia

Australia has one of the fastest deployment rates of renewable energy worldwide. The country has deployed 5.2GW of solar and wind power in 2018 alone and at this rate, is on track to reach 50% renewable electricity in 2024 and 100% in 2032 . However, Australia may be one of the leading major economies in terms of renewable deployments, but it is one of the least prepared at a network level to make this transition, being ranked 28th out of the list of 32 advanced economies on the World Economic Forum's 2019 Energy Transition Index .

Austria

Austria embarked on its energy transition (*Energiewende*) some decades ago. Due to geographical conditions, energy production in Austria relies heavily on renewable energies, notably hydropower. 78.4% of domestic production in 2013 came from renewable energy, 9.2% from natural gas and 7.2% from petroleum. (rest: waste). On the basis of the Federal Constitutional Law for a Nuclear-Free Austria, no nuclear power stations are in operation in Austria. But domestic energy production makes up only 36% of Austria's total energy consumption, which among other things encompasses transport, electricity production, and heating. In 2013, oil accounts for about 36.2% of total energy consumption, renewable energies 29.8%, gas 20.6%, and coal 9.7%. In the past 20 years, the structure of gross domestic energy consumption has shifted from coal and oil to new renewables, in particular between 2005 and 2013 (plus 60%). The EU target for Austria require a renewables share of 34% by 2020 (gross final energy consumption). Austria is on a good way to achieve this target (32.5% in 2013). Energy transition in Austria can be also seen on the local level, in some villages, towns and regions. For example, the town of Güssing in the state of Burgenland is a pioneer in independent and sustainable energy production. Since 2005, Güssing has already produced significantly more heating (58 gigawatt hours) and electricity (14 GWh) from renewable resources than the city itself needs.

Denmark

Denmark, as a country reliant on imported oil, was impacted particularly hard by the 1973 oil crisis. This roused public discussions on building nuclear power stations to diversify energy supply. A strong anti-nuclear movement developed, which fiercely criticized nuclear power plans taken up by the government, and this ultimately led to a 1985 resolution not to build any nuclear power stations in Denmark. The country instead opted for renewable energy, focusing primarily on wind power. Wind turbines for power generation already had a long history in Denmark, as far back as the late 1800s. As early as 1974 a panel of experts declared "that it should be possible to satisfy 10%

of Danish electricity demand with wind power, without causing special technical problems for the public grid." Denmark undertook the development of large wind power stations — though at first with little success (like with the Growian project in Germany).

Small facilities prevailed instead, often sold to private owners such as farms. Government policies promoted their construction; at the same time, positive geographical factors favored their spread, such as good wind power density and Denmark's decentralized patterns of settlement. A lack of administrative obstacles also played a role. Small and robust systems came on line, at first in the power range of only 50-60 kilowatts — using 1940s technology and sometimes hand-crafted by very small businesses. In the late seventies and the eighties a brisk export trade to the United States developed, where wind energy also experienced an early boom. In 1986 Denmark already had about 1200 wind power turbines, though they still accounted for just barely 1% of Denmark's electricity. This share increased significantly over time. In 2011, renewable energies covered 41% of electricity consumption, and wind power facilities alone accounted for 28%. The government aims to increase wind energy's share of power generation to 50% by 2020, while at the same time reducing carbon dioxide emissions by 40%. On 22 March 2012, the Danish Ministry of Climate, Energy and Building published a four-page paper titled "DK Energy Agreement," outlining long-term principles for Danish energy policy.

The installation of oil and gas heating is banned in newly constructed buildings from the start of 2013; beginning in 2016 this will also apply to existing buildings. At the same time an assistance program for heater replacement was launched. Denmark's goal is to reduce the use of fossil fuels 33% by 2020. The country is scheduled to attain complete independence from petroleum and natural gas by 2050.

France

Since 2012, political discussions have been developing in France about the energy transition and how the French economy might profit from it.

In September 2012, Minister of the Environment Delphine Batho coined the term "ecological patriotism." The government began a work plan to consider starting the energy transition in France. This plan should address the following questions by June 2013:

- How can France move towards energy efficiency and energy conservation? Reflections on altered lifestyles, changes in production, consumption, and transport.

- How to achieve the energy mix targeted for 2025? France's climate protection targets call for reducing greenhouse gas emissions 40% by 2030, and 60% by 2040.

- Which renewable energies should France rely on? How should the use of wind and solar energy can be promoted?

- What costs and funding models will likely be required for alternative energy consulting and investment support? And how about for research, renovation, and expansion of district heating, biomass, and geothermal energy? One solution could be a continuation of the CSPE, a tax that is charged on electricity bills.

The Environmental Conference on Sustainable Development on 14 and 15 September 2012 treated the issue of the environmental and energy transition as its main theme.

On 8 July 2013, the national debate leaders submits some proposals to the government. Among them, there were environmental taxation, and smart grid development. In 2015, the National Assembly has adopted legislation for the transition to low emission vehicles.

France is second only to Denmark as having the worlds lowest carbon emissions in relation to gross domestic product.

Germany

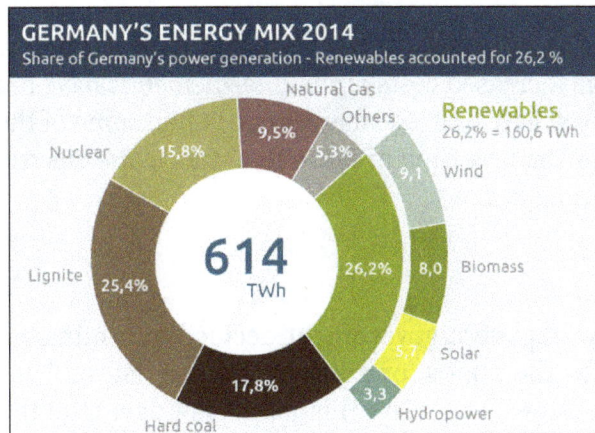

Market share of Germany's power generation 2014

The key policy document outlining the *Energiewende* was published by the German government in September 2010, some six months before the Fukushima nuclear accident. Legislative support was passed in September 2010. Important aspects include:

Key *Energiewende* policy targets (with actual figures for 2014)					
Target	2014	2020	2030	2040	2050
Greenhouse gas emissions					
Greenhouse gas emissions (base year 1990)	−27.0%	−40%	−55%	−70%	−80 to −95%
Renewable energy					
Share of gross final energy consumption	13.5%	18%	30%	45%	60%
Share of gross electricity consumption	27.4%	35%	50%	65%	80%
Efficiency and consumption					
Primary energy consumption (base year 2008)	−8.7%	−20%			−50%
Gross electricity consumption (base year 2008)	−4.6%	−10%			−25%
Final energy consumption in transport (base year 2005)	1.7%	−10%			−40%

In addition, there will be an associated research and development drive.

The policy has been embraced by the German federal government and has resulted in a huge expansion of renewables, particularly wind power. Germanys share of renewables has increased from around 5% in 1999 to 17% in 2010, reaching close to the OECD average of 18% usage of renewables. Producers have been guaranteed a fixed feed-in tariff for 20 years, guaranteeing a fixed income. Energy co-operatives have been created, and efforts were made to decentralize control and profits. The large energy companies have a disproportionately small share of the renewables market. Nuclear power stations were closed, and the existing nine stations will close earlier than necessary, in 2022.

The reduction of reliance on nuclear stations has had the consequence of increased reliance on fossil fuels. One factor that has inhibited efficient employment of new renewable energy has been the lack of an accompanying investment in power infrastructure to bring the power to market. It is believed 8300 km of power lines must be built or upgraded.

Different Länder have varying attitudes to the construction of new power lines. Industry has had their rates frozen and so the increased costs of the *Energiewende* have been passed on to consumers, who have had rising electricity bills. Germans in 2013 had some of the highest electricity costs in Europe. Nonetheless, for the first time in more than ten years, electricity prices for household customers fell at the beginning of 2015.

Japan

On 14 September 2012 the Japanese government decided at a ministerial meeting in Tokyo to phase out nuclear power by the 2030s, or 2040 at the very latest. The government said that it would take "all possible measures" to achieve this goal. A few days later the government retrenched the planned nuclear phaseout after the industry pushed for reconsideration. Arguments cited were that a nuclear phaseout would burden the economy, and that imports of oil, coal, and gas would bring high added costs. The government then approved the energy transition, but left open the time-frame for decommissioning the nuclear power stations.

United Kingdom

The United Kingdom is mainly focusing on wind power, both onshore and offshore, and in particular is strongly promoting the establishment of offshore wind power. With an installed capacity of 18.8 GW at the end of 2017, Britain is one of the worldwide leaders taking the sixth place, after China, the United States, Germany, India, and Spain. It was initially promoted with a quota system, but expansion targets were missed repeatedly. This led the government to implement a feed-in tariff instead.

United States

The Obama administration made a large push for green jobs, particularly in his first term.

In the United States, the share of renewable energy (excluding hydropower) in electricity generation has grown from 3.3 percent (1990) to 5.5 percent (2013). Oil use will decline in the USA owing to the increasing efficiency of the vehicle fleet and replacement of crude oil by natural gas as a feedstock for the petrochemical sector. One forecast is that the rapid uptake of electric vehicles will reduce oil demand drastically, to the point where it is 80% lower in 2050 compared with today.

In December 2016, Block Island Wind Farm became the first commercial US offshore wind farm. It consists of five 6 MW turbines (together 30 MW) located *near-shore* (3.8 miles (6.1 km) from Block Island, Rhode Island) in the Atlantic Ocean. At the same time, Norway-based oil major Statoil laid down nearly $42.5 million on a bid to lease a large offshore area off the coast of New York.

Energy System

An energy system is a system primarily designed to supply energy-services to end-users. Taking a structural viewpoint, the IPCC Fifth Assessment Report defines an energy system as "all components related to the production, conversion, delivery, and use of energy". The field of energy economics includes energy markets and treats an energy system as the technical and economic systems that satisfy consumer demand for energy in the forms of heat, fuels, and electricity.

The first two definitions allow for demand-side measures, including daylighting, retrofitted building insulation, and passive solar building design, as well as socio-economic factors, such as aspects of energy demand management and even telecommuting, while the third does not. Neither does the third account for the informal economy in traditional biomass that is significant in many developing countries.

The analysis of energy systems thus spans the disciplines of engineering and economics. Merging ideas from both areas to form a coherent description, particularly where macroeconomic dynamics are involved, is challenging.

The concept of an energy system is evolving as new regulations, technologies, and practices enter into service – for example, emissions trading, the development of smart grids, and the greater use of energy demand management, respectively.

Treatment

From a structural perspective, an energy system is like any general system and is made up of a set of interacting component parts, located within an environment. These components derive from ideas found in engineering and economics. Taking a process view, an energy system "consists of an integrated set of technical and economic activities operating within a complex societal framework". The identification of the components and behaviors of an energy system depends on the circumstances, the purpose of the analysis, and the questions under investigation. The concept of an energy system is therefore an abstraction which usually precedes some form of computer-based investigation, such as the construction and use of a suitable energy model.

Viewed in engineering terms, an energy system lends itself to representation as a flow network: the vertices map to engineering components like power stations and pipelines and the edges map to the interfaces between these components. This approach allows collections of similar or adjacent components to be aggregated and treated as one to simplify the model. Once described thus, flow network algorithms, such as minimum cost flow, may be applied. The components themselves can be treated as simple dynamical systems in their own right.

Conversely, relatively pure economic modeling may adopt a sectorial approach with only limited engineering detail present. The sector and sub-sector categories published by the International Energy Agency are often used as a basis for this analysis. A 2009 study of the UK residential energy sector contrasts the use of the technology-rich Markal model with several UK sectoral housing stock models.

International energy statistics are typically broken down by carrier, sector and sub-sector, and country. Energy carriers (aka energy products) are further classified as primary energy and secondary (or intermediate) energy and sometimes final (or end-use) energy. Published energy datasets are normally adjusted so that they are internally consistent, meaning that all energy stocks and flows must balance. The IEA regularly publishes energy statistics and energy balances with varying levels of detail and cost and also offers mid-term projections based on this data. The notion of an energy carrier, as used in energy economics, is distinct and different from the definition of energy used in physics.

Energy systems can range in scope, from local, municipal, national, and regional, to global, depending on issues under investigation. Researchers may or may not include demand side measures within their definition of an energy system. The (IPCC) Intergovernmental Panel on Climate Change does so, for instance, but covers these measures in separate chapters on transport, buildings, industry, and agriculture.

Household consumption and investment decisions may also be included within the ambit of an energy system. Such considerations are not common because consumer behavior is difficult to characterize, but the trend is to include human factors in models. Household decision-taking may be represented using techniques from bounded rationality and agent-based behavior. The American Association for the Advancement of Science (AAAS) specifically advocates that "more attention should be paid to incorporating behavioral considerations other than price- and income-driven behavior into economic models of the energy system".

Energy-services

The concept of an energy-service is central, particularly when defining the purpose of an energy system:

> "It is important to realize that the use of energy is no end in itself but is always directed to satisfy human needs and desires. Energy services are the ends for which the energy system provides the means".

Energy-services can be defined as amenities that are either furnished through energy consumption or could have been thus supplied. More explicitly:

> "Demand should, where possible, be defined in terms of energy-service provision, as characterized by an appropriate intensity – for example, air temperature in the case of space-heating or lux levels for illuminance. This approach facilitates a much greater set of potential responses to the question of supply, including the use of energetically-passive techniques – for instance, retrofitted insulation and daylighting".

A consideration of energy-services per capita and how such services contribute to human welfare and individual quality of life is paramount to the debate on sustainable energy. People living in poor regions with low levels of energy-services consumption would clearly benefit from greater consumption, but the same is not generally true for those with high levels of consumption.

The notion of energy-services has given rise to energy-service companies (ESCo) who contract to provide energy-services to a client for an extended period. The ESCo is then free to choose the best means to do so, including investments in the thermal performance and HVAC equipment of the buildings in question.

ISO 13600, 13601, 13602 on Technical Energy Systems

ISO 13600, ISO 13601, and ISO 13602 form a set of international standards covering technical energy systems (TES). Although withdrawn prior to 2016, these documents provide useful definitions and a framework for formalizing such systems. The standards depict an energy system broken down into supply and demand sectors, linked by the flow of tradable energy commodities (or energywares). Each sector has a set of inputs and outputs, some intentional and some harmful byproducts. Sectors may be further divided into subsectors, each fulfilling a dedicated purpose. The demand sector is ultimately present to supply energyware-based services to consumers.

References

- Energy, science: britannica.com, Retrieved 11 January, 2019

- Shinn, E.; et., al. (2012). "Nuclear energy conversion with stacks of graphene nanocapacitors". Complexity. Bibcode:2013Cmplx..18c..24S. Doi:10.1002/cplx.21427

- Conservation-of-energy, science: britannica.com, Retrieved 12 July, 2019

- Nicola Armaroli, Vincenzo Balzani, The Future of Energy Supply: Challenges and Opportunities. In: Angewandte Chemie 46, (2007), 52-66, p. 52, doi:10.1002/anie.200602373

- "Definition of system". Merriam-Webster. Springfield, MA, USA. Retrieved 9 October 2016

Sources of Energy   2

Sources of energy are broadly classified into renewable sources and non-renewable sources. The renewable resources of energy include bioenergy, hydropower and geothermal energy while non-renewable sources of energy are fossil fuels such as coal, natural gas and petroleum. The topics elaborated in this chapter will help in gaining a better perspective about these sources of energy.

Renewable Energy

Renewable energy comes from sources that will not be used up in our lifetimes, such as the sun and wind. The wind, the sun, and Earth are sources of renewable energy. These energy sources naturally renew, or replenish themselves.

Wind, sunlight, and the Earth have energy that transforms in ways we can see and feel. We can see and feel evidence of the transfer of energy from the sun to the Earth in the sunlight shining on the ground and the warmth we feel when sunlight shines on our skin. We can see and feel evidence of the transfer of energy in wind's ability to pull kites higher into the sky and shake the leaves on trees. We can see and feel evidence of the transfer of energy in the geothermal energy of steam vents and geysers.

People have created different ways to capture the energy from these renewable sources.

Solar Energy

Solar energy can be captured "actively" or "passively." Active solar energy uses special technology to capture the sun's rays. The two main types of equipment are photovoltaic cells (also called PV cells or solar cells) and mirrors that focus sunlight in a specific spot. These active solar technologies use sunlight to generate electricity, which we use to power lights, heating systems, computers and televisions.

Passive solar energy does not use any equipment. Instead, it gets energy from the way sunlight naturally changes throughout the day. For example, people can build houses so their windows face the path of the sun. This means the house will get more heat from the sun. It will take less energy from other sources to heat the house.

Other examples of passive solar technology are green roofs, cool roofs, and radiant barriers. Green roofs are completely covered with plants. Plants can get rid of pollutants in rainwater and air. They help make the local environment cleaner.

Cool roofs are painted white. Radiant barriers are made of a reflective covering, such as aluminum. They both reflect the sun's heat instead of absorbing it. All these types of roofs help lower the amount of energy needed to cool the building.

Advantages and Disadvantages of Solar Energy

There are many advantages to using solar energy. PV cells last for a long time, about 20 years. However, there are reasons why solar power cannot be used as the only power source in a community. It can be expensive to install PV cells or build a building using passive solar technology.

Sunshine can also be hard to predict. It can be blocked by clouds, and the sun doesn't shine at night. Different parts of Earth receive different amounts of sunlight based on location, the time of year, and the time of day.

Wind Energy

People have been harnessing the wind's energy for a long, long time. Five thousand years ago, ancient Egyptians made boats powered by the wind. In 200 BCE, people used windmills to grind grain in the Middle East and pump water in China.

Today, we capture the wind's energy with wind turbines. A turbine is similar to a windmill; it has a very tall tower with two or three propeller-like blades at the top. These blades are turned by the wind. The blades turn a generator (located inside the tower), which creates electricity.

Groups of wind turbines are known as wind farms. Wind farms can be found near farmland, in narrow mountain passes, and even in the ocean, where there are steadier and stronger winds. Wind turbines anchored in the ocean are called "offshore wind farms."

Wind farms create electricity for nearby homes, schools, and other buildings.

Advantages and Disadvantages of Wind Energy

Wind energy can be very efficient. In places like the Midwest and along coasts, steady winds can provide cheap, reliable electricity.

Another great advantage of wind power is that it is a "clean" form of energy. Wind turbines do not burn fuel or emit any pollutants into the air.

Wind is not always a steady source of energy, however. Wind speed changes constantly, depending on the time of day, weather, and geographic location. Currently, it cannot be used to provide electricity for all our power needs.

Wind turbines can be also dangerous for bats and birds. These animals cannot always judge how fast the blades are moving and crash into them.

Geothermal Energy

Deep beneath the surface of the Earth is the Earth's core. The center of the Earth is extremely hot—thought to be over 5,000 °C (about 9,000 °F). The heat is constantly moving toward the surface.

We can see some of the Earth's heat when it bubbles to the surface. Geothermal energy can melt underground rocks into magma and cause the magma to bubble to the surface as lava. Geothermal energy can also heat underground sources of water and force it to spew out from the surface. This stream of water is called a geyser. However, most of the Earth's heat stays underground and makes its way out very, very slowly.

In Iceland, there are large reservoirs of underground water. Almost 90% of people in Iceland use geothermal as an energy source to heat their homes and businesses.

Advantages and Disadvantages of Geothermal Energy

An advantage of geothermal energy is that it is clean. It does not require any fuel or emit any harmful pollutants into the air.

A disadvantage of using geothermal energy is that in areas of the world where there is only dry heat underground, large quantities of freshwater are used to make steam. There may not be a lot of freshwater. People need water for drinking, cooking, and bathing.

Biomass Energy

Biomass is any material that comes from plants or microorganisms that were recently living. Plants create energy from the sun through photosynthesis. This energy is stored in the plants even after they die.

Trees, branches, scraps of bark, and recycled paper are common sources of biomass energy. Manure, garbage, and crops such as corn, soy, and sugar cane can also be used as biomass feedstocks.

We get energy from biomass by burning it. Wood chips, manure, and garbage are dried out and compressed into squares called "briquettes." These briquettes are so dry that they do not absorb water. They can be stored and burned to create heat or generate electricity.

Biomass can also be converted into biofuel. Biofuels are mixed with regular gasoline and can be used to power cars and trucks. Biofuels release less harmful pollutants than pure gasoline.

Advantages and Disadvantages of Biomass Energy

A major advantage of biomass is that it can be stored and used when it is needed. Growing crops for biofuels, however, requires large amounts of land and pesticides. Land could be used for food instead of biofuels. Some pesticides could pollute the air and water.

Biomass energy can also be a non-renewable energy source. Biomass energy relies on biomass feedstocks—plants that are processed and burned to create electricity. Biomass feedstocks can include crops such as corn or soy, as well as wood. If people do not replant biomass feedstocks as fast as they use them, biomass energy becomes a non-renewable energy source.

Hydroelectric Energy

Hydroelectric energy is made by flowing water. Most hydroelectric power plants are located on large dams, which control the flow of a river.

Dams block the river and create an artificial lake, or reservoir. A controlled amount of water is forced through tunnels in the dam. As water flows through the tunnels, it turns huge turbines and generates electricity.

Advantages and Disadvantages of Hydroelectric Energy

Hydroelectric energy is fairly inexpensive to harness. Dams do not need to be complex, and the resources to build them are not difficult to obtain. Rivers flow all over the world, so the energy source is available to millions of people.

Hydroelectric energy is also fairly reliable. Engineers control the flow of water through the dam, so the flow does not depend on the weather (the way solar and wind energies do).

However, hydroelectric power plants are damaging to the environment. When a river is dammed, it creates a large lake behind the dam. This lake (sometimes called a reservoir) drowns the original river habitat deep underwater. Sometimes, people build dams that can drown entire towns underwater. The people who live in the town or village must move to a new area.

Hydroelectric power plants don't work for a very long time: Some can only supply power for 20 or 30 years. Silt, or dirt from a riverbed, builds up behind the dam and slows the flow of water.

Other Renewable Energy Sources

Scientists and engineers are constantly working to harness other renewable energy sources. Three of the most promising are tidal energy, wave energy, and algal (or algae) fuel.

Tidal energy harnesses the power of ocean tides to generate electricity. Some tidal energy projects use the moving tides to turn the blades of a turbine. Other projects use small dams to continually fill reservoirs at high tide and slowly release the water (and turn turbines) at low tide.

Wave energy harnesses waves from the ocean, lakes, or rivers. Some wave energy projects use the same equipment that tidal energy projects do—dams and standing turbines. Other wave energy projects float directly on waves. The water's constant movement over and through these floating pieces of equipment turns turbines and creates electricity.

Algal fuel is a type of biomass energy that uses the unique chemicals in seaweed to create a clean and renewable biofuel. Algal fuel does not need the acres of cropland that other biofuel feedstocks do.

This building uses wind turbines, solar panels, and geothermal heat pumps.

Hydropower

Hydropower or water power is power derived from the energy of falling water or fast running water, which may be harnessed for useful purposes. Since ancient times, hydropower from many kinds of watermills has been used as a renewable energy source for irrigation and the operation of various mechanical devices, such as gristmills, sawmills, textile mills, trip hammers, dock cranes, domestic lifts, and ore mills. A trompe, which produces compressed air from falling water, is sometimes used to power other machinery at a distance.

A hydropower scheme which harnesses the power of the water which pours down from the Brecon Beacons mountains.

In the late 19th century, hydropower became a source for generating electricity. Cragside in

Northumberland was the first house powered by hydroelectricity in 1878 and the first commercial hydroelectric power plant was built at Niagara Falls in 1879. In 1881, street lamps in the city of Niagara Falls were powered by hydropower.

Since the early 20th century, the term has been used almost exclusively in conjunction with the modern development of hydroelectric power. International institutions such as the World Bank view hydropower as a means for economic development without adding substantial amounts of carbon to the atmosphere, but dams can have significant negative social and environmental impacts.

Calculating the Amount of Available Power

A hydropower resource can be evaluated by its available power. Power is a function of the hydraulic head and volumetric flow rate. The head is the energy per unit weight (or unit mass) of water. The static head is proportional to the difference in height through which the water falls. Dynamic head is related to the velocity of moving water. Each unit of water can do an amount of work equal to its weight times the head.

The power available from falling water can be calculated from the flow rate and density of water, the height of fall, and the local acceleration due to gravity:

$$\dot{W}_{out} = -\eta\,(\dot{m}g\,\Delta h) = -\eta\,((\rho\dot{V})\,g\,\Delta h)$$

where,

- $\dot{W}_{out}$ (work flow rate out) is the useful power output (in watts).

- η ("eta") is the efficiency of the turbine (dimensionless).

- $\dot{m}$ is the mass flow rate (in kilograms per second).

- ρ ("rho") is the density of water (in kilograms per cubic metre).

- $\dot{V}$ is the volumetric flow rate (in cubic metres per second).

- g is the acceleration due to gravity (in metres per second per second).

- Δh ("Delta h") is the difference in height between the outlet and inlet (in metres).

To illustrate, the power output of a turbine that is 85% efficient, with a flow rate of 80 cubic metres per second (2800 cubic feet per second) and a head of 145 metres (480 feet), is 97 Megawatts:

$$\dot{W}_{out} = 0.85 \times 1000\,(\text{kg}/\text{m}^3) \times 80\,(\text{m}^3/\text{s}) \times 9.81\,(\text{m}/\text{s}^2) \times 145\,\text{m} = 97 \times 10^6\,(\text{kg m}^2/\text{s}^3) = 97\,\text{MW}$$

Operators of hydroelectric stations will compare the total electrical energy produced with the theoretical potential energy of the water passing through the turbine to calculate efficiency. Procedures and definitions for calculation of efficiency are given in test codes such as ASME PTC 18 and IEC 60041. Field testing of turbines is used to validate the manufacturer's guaranteed efficiency. Detailed calculation of the efficiency of a hydropower turbine will account for the head lost due to flow friction in the power canal or penstock, rise in tail water level due to flow, the location of the station and effect of varying gravity, the temperature and barometric pressure of the air, the density of the water at ambient temperature, and the altitudes above sea level of the forebay and tailbay. For precise calculations, errors due to rounding and the number of significant digits of constants must be considered.

Some hydropower systems such as water wheels can draw power from the flow of a body of water without necessarily changing its height. In this case, the available power is the kinetic energy of the flowing water. Over-shot water wheels can efficiently capture both types of energy. The water flow in a stream can vary widely from season to season. Development of a hydropower site requires analysis of flow records, sometimes spanning decades, to assess the reliable annual energy supply. Dams and reservoirs provide a more dependable source of power by smoothing seasonal changes in water flow. However reservoirs have significant environmental impact, as does alteration of naturally occurring stream flow. The design of dams must also account for the worst-case, "probable maximum flood" that can be expected at the site; a spillway is often included to bypass flood flows around the dam. A computer model of the hydraulic basin and rainfall and snowfall records are used to predict the maximum flood.

Mechanical Power

Compressed Air Hydro

Where there is a plentiful head of water it can be made to generate compressed air directly without moving parts. In these designs, a falling column of water is purposely mixed with air bubbles generated through turbulence or a venturi pressure reducer at the high-level intake. This is allowed to fall down a shaft into a subterranean, high-roofed chamber where the now-compressed air separates from the water and becomes trapped. The height of the falling water column maintains compression of the air in the top of the chamber, while an outlet, submerged below the water level in the chamber allows water to flow back to the surface at a lower level than the intake. A separate outlet in the roof of the chamber supplies the compressed air. A facility on this principle was built on the Montreal River at Ragged Shutes near Cobalt, Ontario in 1910 and supplied 5,000 horsepower to nearby mines.

A shishi-odoshi powered by falling water breaks the quietness of a Japanese garden with the sound of a bamboo rocker arm hitting a rock.

Hydroelectricity

Hydroelectricity is the application of hydropower to generate electricity. It is the primary use of hydropower today. Hydroelectric power plants can include a reservoir (generally created by a dam) to exploit the energy of falling water, or can use the kinetic energy of water as in run-of-the-river hydroelectricity. Hydroelectric plants can vary in size from small community sized plants (micro hydro) to very large plants supplying power to a whole country. As of 2019, the five largest power stations in the world are conventional hydroelectric power stations with dams.

Hydroelectricity can also be used to store energy in the form of potential energy between two reservoirs at different heights with pumped-storage hydroelectricity. Water is pumped uphill into reservoirs during periods of low demand to be released for generation when demand is high or system generation is low.

Other forms of electricity generation with hydropower include tidal stream generators using energy from tidal power generated from oceans, rivers, and human-made canal systems to generating electricity.

Geothermal Energy

Geothermal energy is a form of energy conversion in which heat energy from within Earth is captured and harnessed for cooking, bathing, space heating, electrical power generation, and other uses.

Heat from Earth's interior generates surface phenomena such as lava flows, geysers, fumaroles, hot springs, and mud pots. The heat is produced mainly by the radioactive decay of potassium, thorium, and uranium in Earth's crust and mantle and also by friction generated along the margins of continental plates. The subsequent annual low-grade heat flow to the surface averages between 50 and 70 milliwatts (mW) per square metre worldwide. In contrast, incoming solar radiation striking Earth's surface provides 342 watts per square metre annually. Geothermal heat energy can be recovered and exploited for human use, and it is available anywhere on Earth's surface. The estimated energy that can be recovered and utilized on the surface is 4.5×10^6 exajoules, or about 1.4×10^6 terawatt-years, which equates to roughly three times the world's annual consumption of all types of energy.

The amount of usable energy from geothermal sources varies with depth and by extraction method. The increase in temperature of rocks and other materials underground averages 20–30 °C (36–54 °F) per kilometre (0.6 mile) depth worldwide in the upper part of the lithosphere, and this rate of increase is much higher in most of Earth's known geothermal areas. Normally, heat extraction requires a fluid (or steam) to bring the energy to the surface. Locating and developing geothermal resources can be challenging. This is especially true for the high-temperature resources needed for generating electricity. Such resources are typically limited to parts of the world characterized by recent volcanic activity or located along plate boundaries or within crustal hot spots. Even though there is a continuous source of heat within Earth, the extraction rate of the heated fluids and steam can exceed the replenishment rate, and, thus, use of the resource must be managed sustainably.

Uses

Geothermal energy use can be divided into three categories: direct-use applications, geothermal heat pumps (GHPs), and electric power generation.

Direct Uses

Probably the most widely used set of applications involves the direct use of heated water from the ground without the need for any specialized equipment. All direct-use applications make use of low-temperature geothermal resources, which range between about 50 and 150 °C (122 and 302 °F). Such low-temperature geothermal water and steam have been used to warm single buildings, as well as whole districts where numerous buildings are heated from a central supply source. In addition, many swimming pools, balneological (therapeutic) facilities at spas, greenhouses, and aquaculture ponds around the world have been heated with geothermal resources. Other direct uses of geothermal energy include cooking, industrial applications (such as drying fruit, vegetables, and timber), milk pasteurization, and large-scale snow melting. For many of those activities, hot water is often used directly in the heating system, or it may be used in conjunction with a heat exchanger, which transfers heat when there are problematic minerals and gases such as hydrogen sulfide mixed in with the fluid.

Hot springs in Bagno Vignoni.

Geothermal Heat Pumps

Geothermal heat pumps (GHPs) take advantage of the relatively stable moderate temperature conditions that occur within the first 300 metres (1,000 feet) of the surface to heat buildings in the winter and cool them in the summer. In that part of the lithosphere, rocks and groundwater occur at temperatures between 5 and 30 °C (41 and 86 °F). At shallower depths, where most GHPs are found, such as within 6 metres (about 20 feet) of Earth's surface, the temperature of the ground maintains a near-constant temperature of 10 to 16 °C (50 to 60 °F). Consequently, that heat can be used to help warm buildings during the colder months of the year when the air temperature falls below that of the ground. Similarly, during the warmer months of the year, warm air can be drawn from a building and circulated underground, where it loses much of its heat and is returned.

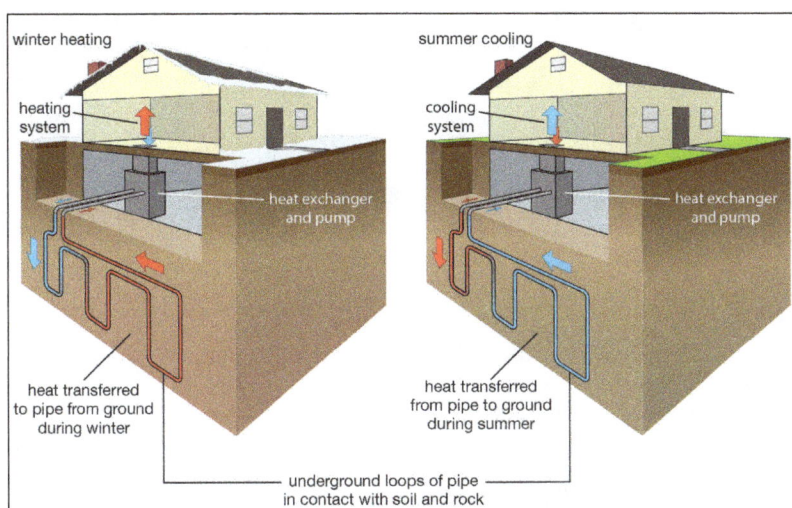

Residential heat pump operation for summer cooling and winter heating.

A GHP system is made up of a heat exchanger (a loop of pipes buried in the ground) and a pump. The heat exchanger transfers heat energy between the ground and air at the surface by means of a fluid that circulates through the pipes; the fluid used is often water or a combination of water and antifreeze. During warmer months, heat from warm air is transferred to the heat exchanger and into the fluid. As it moves through the pipes, the heat is dispersed to the rocks, soil, and groundwater. The pump is reversed during the colder months. Heat energy stored in the relatively warm ground raises the temperature of the fluid. The fluid then transfers this energy to the heat pump, which warms the air inside the building.

GHPs have several advantages over more conventional heating and air-conditioning systems. They are very efficient, using 25–50 percent less electricity than comparable conventional heating and cooling systems, and they produce less pollution. The reduction in energy use associated with GHPs can translate into as much as a 44 percent decrease in greenhouse gas emissions compared with air-source heat pumps (which transfer heat between indoor and outdoor air). In addition, when compared with electric resistance heating systems (which convert electricity to heat) coupled with standard air-conditioning systems, GHPs can produce up to 72 percent less greenhouse gas emissions.

Electric Power Generation

Depending upon the temperature and the fluid (steam) flow, geothermal energy can be used to generate electricity. Geothermal power plants can produce electricity in three ways. Despite their differences in design, all three control the behaviour of steam and use it to drive electrical generators. Given that the excess water vapour at the end of each process is condensed and returned to the ground, where it is reheated for later use, geothermal power is considered a form of renewable energy.

Some geothermal power plants simply collect rising steam from the ground. In such "dry steam" operations, the heated water vapour is funneled directly into a turbine that drives an electrical generator. Other power plants, built around the flash steam and binary cycle designs, use a mixture of steam and heated water ("wet steam") extracted from the ground to start the electrical generation process.

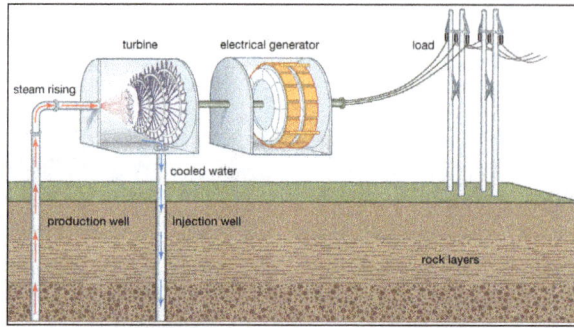

Dry steam geothermal power generation.

In flash steam power plants, pressurized high-temperature water is drawn from beneath the surface into containers at the surface, called flash tanks, where the sudden decrease in pressure causes the liquid water to "flash," or vaporize, into steam. The steam is then used to power the turbine-generator set. In contrast, binary-cycle power plants use steam driven off a secondary working fluid (such as ammonia and hydrocarbons) contained within a closed loop of pipes to power the turbine-generator set. In this process, geothermally heated water is drawn up through a different set of pipes, and much of the energy stored in the heated water is transferred to the working fluid through a heat exchanger. The working fluid then vaporizes. After the vapour from the working fluid passes through the turbine, it is recondensed and piped back to the heat exchanger.

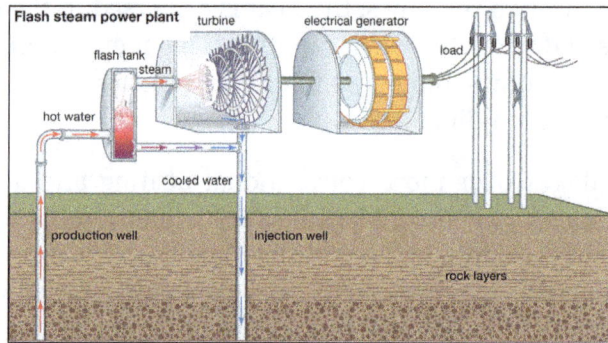

Flash steam geothermal power generation.

Electrical power usually requires water heated above 175 °C (347 °F) to be economical. In geothermal plants using the Organic Rankine Cycle (ORC), a special type of binary-cycle technology that utilizes lower-temperature heat sources (such as biomass combustion and industrial waste heat), water temperatures as low as 85–90 °C (185–194 °F) may be used.

Extraction

Geothermal energy is best found in areas with high thermal gradients. Those gradients occur in regions affected by recent volcanism, in areas located along plate boundaries (such as along the Pacific Ring of Fire), or in areas marked by thin crust (hot spots) such as Yellowstone National Park and the Hawaiian Islands. Geothermal reservoirs associated with those regions must have a heat source, adequate water recharge, a reservoir with adequate permeability or faults that allow fluids to rise close to the surface, and an impermeable caprock to prevent the escape of the heat. In addition, such reservoirs must be economically accessible (that is, within the range of drills).

A geothermal power station in Iceland that creates electricity from heat generated in Earth's interior.

The heated fluid from a geothermal resource is tapped by drilling wells, sometimes as deep as 9,100 metres (about 30,000 feet), and is extracted by pumping or by natural artesian flow (where the weight of the water forces it to the surface). Water and steam are then piped to the power plant to generate electricity or through insulated pipelines—which may be buried or placed aboveground— for use in heating and cooling applications. In general, electric power plant pipelines are limited to roughly 1.6 km (1 mile) in length to minimize heat loss in the steam. However, direct-use pipelines spanning several tens of kilometres have been installed with a temperature loss of less than 2–5 °C (3.6–9 °F), depending on the flow rate. The most economically efficient facilities are located close to the geothermal resource to minimize the expense of constructing long pipelines. In the case of electric power generation, costs can be kept down by locating the facility near electrical transmission lines to transmit the electricity to market.

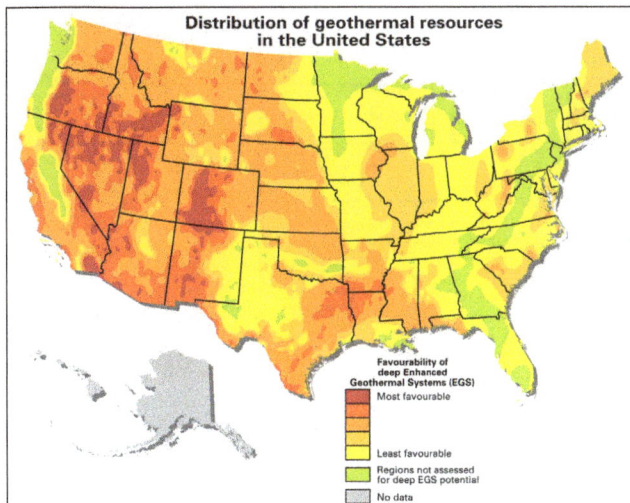

Map of geothermal energy resources in the United States.

Exhaustion

Geothermal resources can be exhausted if the rate of heat extraction exceeds the rate of natural heat recharge. Normally, geothermal resources can be used for 20 to 30 years; however, the energy output may decrease with time, making continued development uneconomical. On the other hand, geothermal electric power has been produced continually from the Larderello geothermal field since the early 1900s and at the Geysers since 1960. Although there has been a decline in both of those fields, this problem has been partially overcome by drilling new wells and by recharging the water supply. At the Geysers, electrical capacity declined from 1,800 MW to approximately 1,000

MW, but about 200 MW of capacity was returned by placing the field under one operator and constructing pipelines to deliver wastewater for recharging the reservoir. Projects such as the Reykjavík district heating system have been operating since the 1930s with little change in the output, and the Oregon Institute of Technology geothermal heating system has been operating since the 1950s with no change in production. Thus, with proper management, geothermal resources can be sustainable for many years, and they can even recover if use is suspended for a period of time.

Environmental Effects and Economic Costs

The environmental effects of geothermal development and power generation include the changes in land use associated with exploration and plant construction, noise and sight pollution, the discharge of water and gases, the production of foul odours, and soil subsidence. Most of those effects, however, can be mitigated with current technology so that geothermal uses have no more than a minimal impact on the environment. For example, Klamath Falls, Oregon, has approximately 600 geothermal wells for residential space heating. The city has also invested in a district heating system and a downtown snow-melting system, and it provides heating to local businesses. However, none of the systems used to supply and deliver geothermal energy are visible in town.

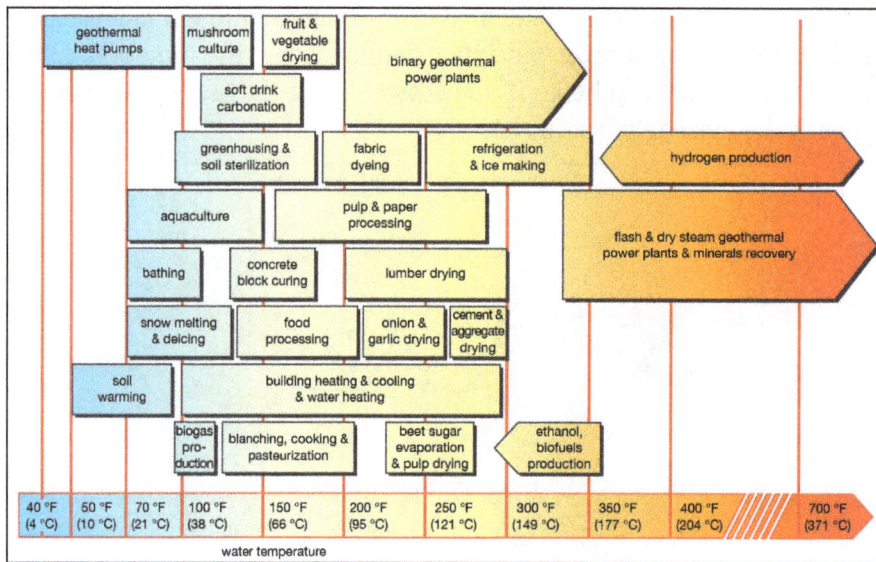

Geothermal energy uses.

Above is shown diagram of various geothermal energy uses displayed according to the water temperature of the geothermal resource.

In addition, GHPs have a very minimal effect on the environment, because they make use of shallow geothermal resources within 100 metres (about 330 feet) of the surface. GHPs cause only small temperature changes to the groundwater or rocks and soil in the ground. In closed-loop systems the ground temperature around the vertical boreholes is slightly increased or decreased; the direction of the temperature change is governed by whether the system is dominated by heating (which would be the case in colder regions) or cooling (which would be the case in warmer regions). With balanced heating and cooling loads, the ground temperatures will remain stable. Likewise, open-loop systems using groundwater or lake water would have very little effect on temperature, especially in regions characterized by high groundwater flows.

Comparing the benefits of geothermal energy with other renewable energy sources, the main advantage of geothermal energy is that its base load is available 24 hours per day, 7 days per week, whereas solar and wind are available only about one-third of the time. In addition, the cost of geothermal energy varies between 5 and 10 cents per kilowatt-hour, which can be competitive with other energy sources, such as coal. The main disadvantage of geothermal energy development is the high initial investment cost in constructing the facilities and infrastructure and the high risk of proving the resources. (Geothermal resources in low-permeability rocks are often found, and exploration activities often drill "dry" holes—that is, holes that produce steam in amounts too low to be exploited economically.) However, once the resource is proven, the annual cost of fuel (that is, hot water and steam) is low and tends not to escalate in price.

Bioenergy

Bioenergy is renewable energy made available from materials derived from biological sources. Biomass is any organic material which has stored sunlight in the form of chemical energy. As a fuel it may include wood, wood waste, straw, and other crop residues , manure, sugarcane, and many other by-products from a variety of agricultural processes. By 2010, there was 35 GW (47,000,000 hp) of globally installed bioenergy capacity for electricity generation, of which 7 GW (9,400,000 hp) was in the United States.

In its most narrow sense it is a synonym to biofuel, which is fuel derived from biological sources. In its broader sense it includes biomass, the biological material used as a biofuel, as well as the social, economic, scientific and technical fields associated with using biological sources for energy. This is a common misconception, as bioenergy is the energy extracted from the biomass, as the biomass is the fuel and the bioenergy is the energy contained in the fuel.

There is a slight tendency for the word *bioenergy* to be favoured in Europe compared with *biofuel* in America.

Solid Biomass

Simple use of biomass fuel
(Combustion of wood for heat).

Sugarcane plantation to produce ethanol.

One of the advantages of biomass fuel is that it is often a by-product, residue or waste-product of other processes, such as farming, animal husbandry and forestry. In theory this means there is no competition between fuel and food production, although this is not always the case. Land use, existing biomass industries and relevant conversion technologies must be considered when evaluating suitability of developing biomass as feedstock for energy.

A CHP power station using wood to supply 30,000 households.

Biomass is the material derived from recently living organisms, which includes plants, animals and their byproducts. Manure, garden waste and crop residues are all sources of biomass. It is a renewable energy source based on the carbon cycle, unlike other natural resources such as petroleum, coal, and nuclear fuels. Another source includes Animal waste, which is a persistent and unavoidable pollutant produced primarily by the animals housed in industrial-sized farms.

There are also agricultural products specifically being grown for biofuel production. These include corn, and soybeans and to some extent willow and switchgrass on a pre-commercial research level, primarily in the United States; rapeseed, wheat, sugar beet, and willow (15,000 ha or 37,000 acres in Sweden) primarily in Europe; sugarcane in Brazil; palm oil and miscanthus in Southeast Asia; sorghum and cassava in China; and jatropha in India. Hemp has also been proven to work as a biofuel. Biodegradable outputs from industry, agriculture, forestry and households can be used for biofuel production, using e.g. anaerobic digestion to produce biogas, gasification to produce syngas or by direct combustion. Examples of biodegradable wastes include straw, timber, manure, rice husks, sewage, and food waste. The use of biomass fuels can therefore contribute to waste management as well as fuel security and help to prevent or slow down climate change, although alone they are not a comprehensive solution to these problems.

Biomass can be converted to other usable forms of energy like methane gas or transportation fuels like ethanol and biodiesel. Rotting garbage, and agricultural and human waste, all release methane gas—also called "landfill gas" or "biogas." Crops, such as corn and sugar cane, can be fermented to produce the transportation fuel, ethanol. Biodiesel, another transportation fuel, can be produced from left-over food products like vegetable oils and animal fats. Also, Biomass to liquids (BTLs) and cellulosic ethanol are still under research.

Sewage Biomass

The use of municipal and household waste is on the forefront of new sources for biomass, and is a largely discarded resource on which new research is being conducted for use of energy production. A new bioenergy sewage treatment process aimed at developing countries is now on the horizon;

the Omni Processor is a self-sustaining process which uses the sewerage solids as fuel to convert sewage waste water into drinking water and electrical energy. Sewage sludge is a point of focus in current research for developing bioenergy from biomass. The large quantity being produced by households at a continuous rate presents an opportunity to extract valuable compounds contained within it which can be then used to produce bioenergy. The main form of bioenergy being produced from sewage is methane, but producing other forms is still being researched. The use of sewage to produce methane reduces the amount of waste put into landfills, its costs of transportation and disposal, and also keeps a larger amount of gas out of the atmosphere, as more is able to be captured.

Electricity Generation from Biomass

The biomass used for electricity production ranges by region. Forest byproducts, such as wood residues, are popular in the United States. Agricultural waste is common in Mauritius (sugar cane residue) and Southeast Asia (rice husks). Animal husbandry residues, such as poultry litter, is popular in the UK.

Electricity from Sugarcane Bagasse

Sugarcane (*Saccharum officinarum*) plantation ready for harvest.

A sugar/ethanol plant located in Piracicaba, São Paulo State. This plant produces the electricity
it needs from bagasse residuals from sugarcane left over by the milling process,
and it sells the surplus electricity to the public grid.

Sucrose accounts for little more than 30% of the chemical energy stored in the mature plant; 35% is in the leaves and stem tips, which are left in the fields during harvest, and 35% are in the fibrous material (bagasse) left over from pressing.

The production process of sugar and ethanol in takes full advantage of the energy stored in sugarcane. Part of the bagasse is currently burned at the mill to provide heat for distillation and electricity to run the machinery. This allows ethanol plants to be energetically self-sufficient and even sell surplus electricity to utilities; current production is 600 MW (800,000 hp) for self-use and 100 MW (130,000 hp) for sale. This secondary activity is expected to boom now that utilities have been induced to pay "fair price "(about US$10/GJ or US$0.036/kWh) for 10 year contracts. This is approximately half of what the World Bank considers the reference price for investing in similar projects. The energy is especially valuable to utilities because it is produced mainly in the dry season when hydroelectric dams are running low. Estimates of potential power generation from bagasse range from 1,000 to 9,000 MW (1,300,000 to 12,100,000 hp), depending on technology. Higher estimates assume gasification of biomass, replacement of current low-pressure steam boilers and turbines by high-pressure ones, and use of harvest trash currently left behind in the fields. For comparison, Brazil's Angra I nuclear plant generates 657 MW (881,000 hp).

Presently, it is economically viable to extract about 288 MJ of electricity from the residues of one tonne of sugarcane, of which about 180 MJ are used in the plant itself. Thus a medium-size distillery processing 1,000,000 tonnes (980,000 long tons; 1,100,000 short tons) of sugarcane per year could sell about 5 MW (6,700 hp) of surplus electricity. At current prices, it would earn US$18 million from sugar and ethanol sales, and about US$1 million from surplus electricity sales. With advanced boiler and turbine technology, the electricity yield could be increased to 648 MJ per tonne of sugarcane, but current electricity prices do not justify the necessary investment. (According to one report, the World Bank would only finance investments in bagasse power generation if the price were at least US$19/GJ or US$0.068/kWh.)

Bagasse burning is environmentally friendly compared to other fuels like oil and coal. Its ash content is only 2.5% (against 30–50% of coal), and it contains very little sulfur. Since it burns at relatively low temperatures, it produces little nitrous oxides. Moreover, bagasse is being sold for use as a fuel (replacing heavy fuel oil) in various industries, including citrus juice concentrate, vegetable oil, ceramics, and Tyre Recycling. The state of São Paulo alone used 2,000,000 tonnes (1,970,000 long tons; 2,200,000 short tons), saving about US$35 million in fuel oil imports.

Electricity from Electrogenic Micro-organisms

Another form of bioenergy can be attained from microbial fuel cells, in which chemical energy stored in wastewater or soil is converted directly into electrical energy via the metabolic processes of electrogenic micro-organisms. The power generation capability of this technology has not been found to be economically viable till date, however, this technology has found been found to be more useful for chemical treatment processes and student education.

Environmental Impact

Some forms of forest bioenergy have recently come under fire from a number of environmental organizations, including Greenpeace and the Natural Resources Defense Council, for the harmful impacts they can have on forests and the climate. Greenpeace recently released a report entitled Fuelling a BioMess which outlines their concerns around forest bioenergy. Because any part of the tree can be burned, the harvesting of trees for energy production encourages Whole-Tree Harvesting, which removes more nutrients and soil cover than regular harvesting, and can be harmful to the long-term health of the forest. In some jurisdictions, forest biomass is increasingly consisting of elements essential to functioning forest ecosystems, including standing trees, naturally disturbed forests and remains of traditional logging operations that were previously left in the forest. Environmental groups also cite recent scientific research which has found that it can take many decades for the carbon released by burning biomass to be recaptured by regrowing trees, and even longer in low productivity areas; furthermore, logging operations may disturb forest soils and cause them to release stored carbon. In light of the pressing need to reduce greenhouse gas emissions in the short term in order to mitigate the effects of climate change, a number of environmental groups are opposing the large-scale use of forest biomass in energy production.

Bioenergy can be bad: Suppose you cut down a 50-year oak tree in your garden and use the logs to heat your house instead of coal. Wood emits more carbon dioxide than coal per unit of heat gained and the roots left in the soil emit more carbon dioxide as they rot. If you plant another tree it will soak up that carbon dioxide in about 50 years. But if you had left the original tree in place it would have soaked up the carbon dioxide from the coal and more. It could take centuries before cutting down the tree would give any benefit. But the world needed to cut carbon dioxide over the next few decades if the global warming was to be kept below 3 degrees C. It concluded that official claimed carbon reductions from renewables had been overstated. The European Union, for example, got more 64% of its renewable energy from biomass (mostly wood) but United Nations and EU rules did not count the carbon emissions from burning biomass.

Recently, a new company called Mango materials used bacterial fermentation to produce an intracellular biopolymer, polyhydroxyalkanoate from methane. The great advantage of biopolymers is that it is biodegradable which makes it environment friendly. Because methane is being used that decreases the price of polymers that it would compete with traditional plastics. Also, because methane would be converted into biopolymer that would reduce methane emissions. Chief Executive Officer Molly Morse said that the unused methane would be enough to produce more than three billion pounds of biopolymer. Morse announced in 2017 that using this polymer will reduce the waste in the textile industry because it will be reproduced as biopolymer again in every effective industrial loop.

Non-renewable Energy

Non-renewable energy comes from sources that will run out or will not be replenished in our lifetimes—or even in many, many lifetimes.

Most non-renewable energy sources are fossil fuels: coal, petroleum, and natural gas. Carbon is the main element in fossil fuels. For this reason, the time period that fossil fuels formed (about 360-300 million years ago) is called the Carboniferous Period.

All fossil fuels formed in a similar way. Hundreds of millions of years ago, even before the dinosaurs, Earth had a different landscape. It was covered with wide, shallow seas and swampy forests.

Plants, algae, and plankton grew in these ancient wetlands. They absorbed sunlight and created energy through photosynthesis. When they died, the organisms drifted to the bottom of the sea or lake. There was energy stored in the plants and animals when they died.

Over time, the dead plants were crushed under the seabed. Rocks and other sediment piled on top of them, creating high heat and pressure underground. In this environment, the plant and animal remains eventually turned into fossil fuels (coal, natural gas, and petroleum). Today, there are huge underground pockets (called reservoirs) of these non-renewable sources of energy all over the world.

Advantages and Disadvantages

Fossil fuels are a valuable source of energy. They are relatively inexpensive to extract. They can also be stored, piped, or shipped anywhere in the world.

However, burning fossil fuels is harmful for the environment. When coal and oil are burned, they release particles that can pollute the air, water, and land. Some of these particles are caught and set aside, but many of them are released into the air.

Burning fossil fuels also upsets Earth's "carbon budget," which balances the carbon in the ocean, earth, and air. When fossil fuels are combusted (heated), they release carbon dioxide into the atmosphere. Carbon dioxide is a gas that keeps heat in Earth's atmosphere, a process called the "greenhouse effect." The greenhouse effect is necessary to life on Earth, but relies on a balanced carbon budget.

The carbon in fossil fuels has been sequestered, or stored, underground for millions of years. By removing this sequestered carbon from the earth and releasing it into the atmosphere, Earth's carbon budget is out of balance. This contributes to temperatures rising faster than organisms can adapt.

Coal

Coal is a black or brownish rock. We burn coal to create energy. Coal is ranked depending on how much "carbonization" it has gone through. Carbonization is the process that ancient organisms undergo to become coal. About 3 meters (10 feet) of solid vegetation crushed together into .3 meter (1 foot) of coal.

Advantages and Disadvantages of Coal

Coal is a reliable source of energy. We can rely on it day and night, summer and winter, sunshine or rain, to provide fuel and electricity.

Using coal is also harmful. Mining is one of the most dangerous jobs in the world. Coal miners are exposed to toxic dust and face the dangers of cave-ins and explosions at work.

When coal is burned, it releases many toxic gases and pollutants into the atmosphere. Mining for coal can also cause the ground to cave in and create underground fires that burn for decades at a time.

Petroleum

Petroleum is a liquid fossil fuel. It is also called oil or crude oil. Petroleum is trapped by underground rock formations. In some places, oil bubbles right out of the ground. At the LaBrea Tar Pits, in Los Angeles, California, big pools of thick oil bubble up through the ground. Remains of animals that got trapped there thousands of years ago are still preserved in the tar.

Advantages and Disadvantages of Petroleum

There are advantages to drilling for oil. It is relatively inexpensive to extract. It is also a reliable and dependable source of energy and money for the local community.

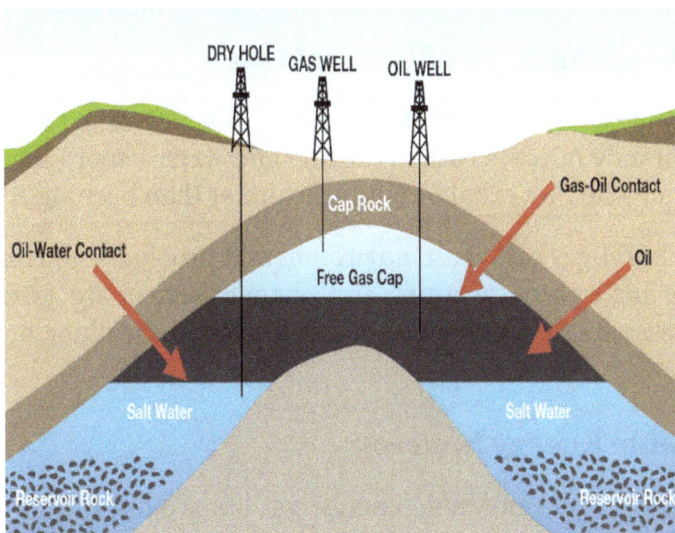

Oil provides us with thousands of conveniences. In the form of gasoline, it is a portable source of energy that gives us the power to drive places. Petroleum is also an ingredient in many items that we depend on.

However, burning gasoline is harmful to the environment. It releases hazardous gases and fumes into the air that we breathe. There is also the possibility of an oil spill. If there is a problem with the drilling machinery, the oil can explode out of the well and spill into the ocean or surrounding land. Oil spills are environmental disasters, especially offshore spills. Oil floats on water, so it can look like food to fish and ruin birds' feathers.

Natural Gas

Natural gas is another fossil fuel that is trapped underground in reservoirs. It is mostly made up of methane. You may have smelled methane before. The decomposing material in landfills also release methane, which smells like rotten eggs.

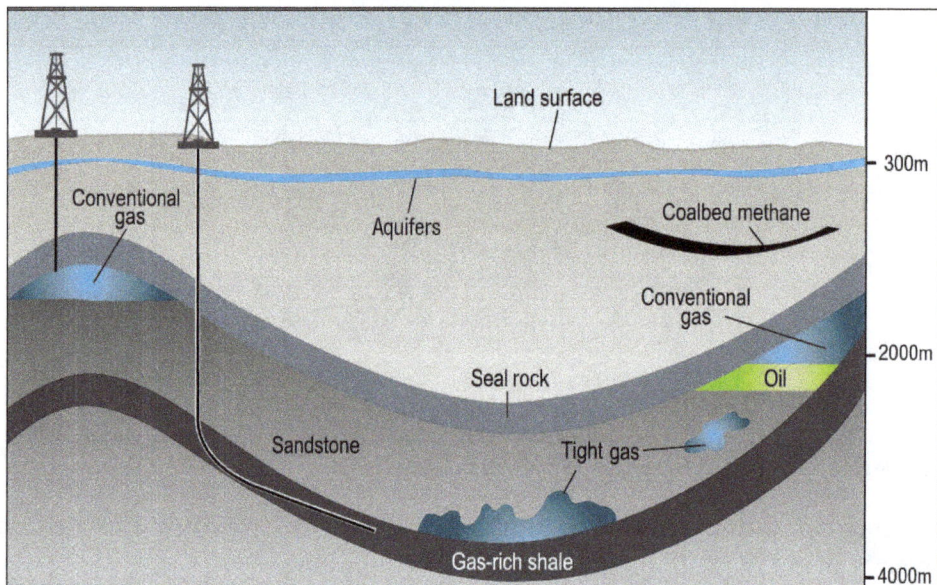

Advantages and Disadvantages of Natural Gas

Natural gas is relatively inexpensive to extract, and is a "cleaner" fossil fuel than oil or coal. When natural gas is burned, it only releases carbon dioxide and water vapor (which are the exact same gases that we breathe out when we exhale.) This is healthier than burning coal.

However, extracting natural gas can cause environmental problems. Fracturing rocks can cause mini-earthquakes. The high-pressure water and chemicals that are forced underground can also leak to other sources of water. The water sources, used for drinking or bathing, can become contaminated and unsafe.

Other Non-renewable Energy Sources

Fossil fuels are the leading non-renewable energy sources around the world. There are others, however.

Nuclear Energy

Nuclear energy is usually considered another non-renewable energy source. Although nuclear energy itself is a renewable energy source, the material used in nuclear power plants is not.

Nuclear energy harvests the powerful energy in the nucleus, or core, of an atom. Nuclear energy is released through nuclear fission, the process where the nucleus of an atom splits. Nuclear power plants are complex machines that can control nuclear fission to produce electricity.

The material most often used in nuclear power plants is the element uranium. Although uranium is found in rocks all over the world, nuclear power plants usually use a very rare type of uranium, U-235. Uranium is a non-renewable resource.

Nuclear energy is a popular way of generating electricity around the world. Nuclear power plants do not pollute the air or emit greenhouse gases. They can be built in rural or urban areas, and do not destroy the environment around them.

However, nuclear energy is difficult to harvest. Nuclear power plants are very complicated to build and run. Many communities do not have the scientists and engineers to develop a safe and reliable nuclear energy program.

Nuclear energy also produces radioactive material. Radioactive waste can be extremely toxic, causing burns and increasing the risk for cancers, blood diseases, and bone decay among people who are exposed to it.

Biomass Energy

Natural gas is one non-renewable energy source.

Biomass energy, a renewable energy source, can also be a non-renewable energy source. Biomass energy uses the energy found in plants.

Biomass energy relies on biomass feedstocks—plants that are processed and burned to create electricity. Biomass feedstocks can include crops such as corn or soy, as well as wood. If people do not replant biomass feedstocks as fast as they use them, biomass energy becomes a non-renewable energy source.

Fossil Fuels

Fossil fuel refers to any of a class of hydrocarbon-containing materials of biological origin occurring within Earth's crust that can be used as a source of energy.

Fossil fuel: Coal is burned to fuel this electric power plant in Rock Springs.

Fossil fuels include coal, petroleum, natural gas, oil shales, bitumens, tar sands, and heavy oils. All contain carbon and were formed as a result of geologic processes acting on the remains of organic matter produced by photosynthesis, a process that began in the Archean Eon (4.0 billion to 2.5 billion years ago). Most carbonaceous material occurring before the Devonian Period (419.2 million to 358.9 million years ago) was derived from algae and bacteria, whereas most carbonaceous material occurring during and after that interval was derived from plants.

All fossil fuels can be burned in air or with oxygen derived from air to provide heat. This heat may be employed directly, as in the case of home furnaces, or used to produce steam to drive generators that can supply electricity. In still other cases—for example, gas turbines used in jet aircraft—the heat yielded by burning a fossil fuel serves to increase both the pressure and the temperature of the combustion products to furnish motive power.

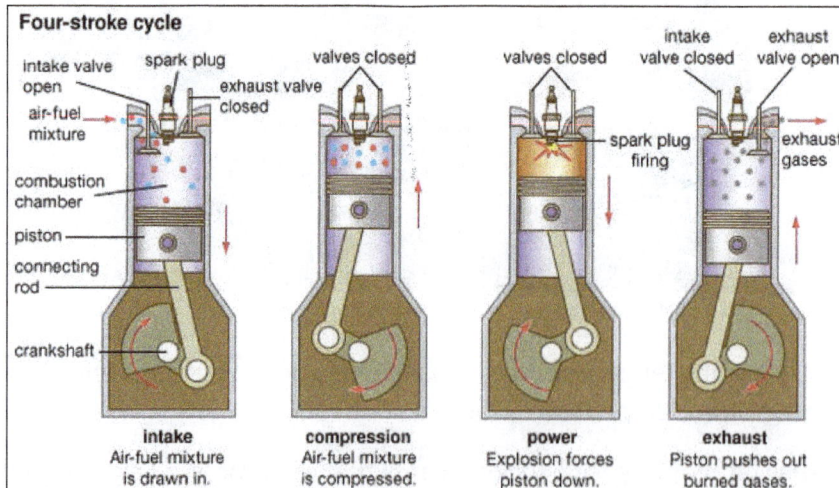

Internal-combustion engine: four-stroke cycle-An internal-combustion engine goes through four strokes: intake, compression, combustion (power), and exhaust. As the piston moves during each stroke, it turns the crankshaft.

Since the beginning of the Industrial Revolution in Great Britain in the second half of the 18th century, fossil fuels have been consumed at an ever-increasing rate. Today they supply more than 80 percent of all the energy consumed by the industrially developed countries of the world. Although new deposits continue to be discovered, the reserves of the principal fossil fuels remaining on Earth are limited. The amounts of fossil fuels that can be recovered economically are difficult to estimate, largely because of changing rates of consumption and future value as well as technological developments. Advances in technology—such as hydraulic fracturing (fracking), rotary drilling, and directional drilling—have made it possible to extract smaller and difficult-to-obtain deposits of fossil fuels at a reasonable cost, thereby increasing the amount of recoverable material. In addition, as recoverable supplies of conventional (light-to-medium) oil became depleted, some petroleum-producing companies shifted to extracting heavy oil, as well as liquid petroleum pulled from tar sands and oil shales.

One of the main by-products of fossil fuel combustion is carbon dioxide (CO_2). The ever-increasing use of fossil fuels in industry, transportation, and construction has added large amounts of CO_2 to Earth's atmosphere. Atmospheric CO_2 concentrations fluctuated between 275 and 290 parts per million by volume (ppmv) of dry air between 1000 CE and the late 18th century but increased to 316 ppmv by 1959 and rose to 412 ppmv in 2018. CO_2 behaves as a greenhouse gas—that is, it absorbs infrared radiation (net heat energy) emitted from Earth's surface and reradiates it back to the surface. Thus, the substantial CO_2 increase in the atmosphere is a major contributing factor to human-induced global warming. Methane (CH_4), another potent greenhouse gas, is the chief constituent of natural gas, and CH_4 concentrations in Earth's atmosphere rose from 722 parts per billion (ppb) before 1750 to 1,859 ppb by 2018. To counter worries over rising greenhouse gas concentrations and to diversify their energy mix, many countries have sought to reduce their dependence on fossil fuels by developing sources of renewable energy (such as wind, solar, hydroelectric, tidal, geothermal, and biofuels) while at the same time increasing the mechanical efficiency of engines and other technologies that rely on fossil fuels.

Coal

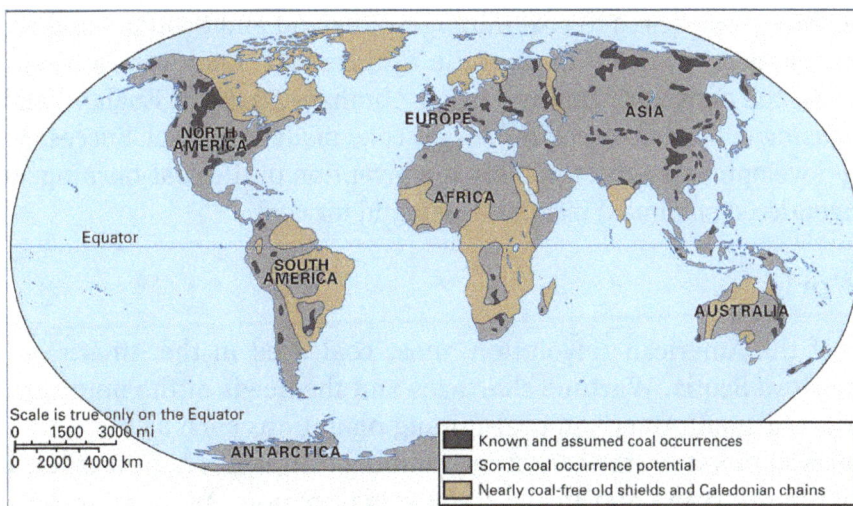

Location of the most-important coal occurrences on Earth.

Coal is one of the most important primary fossil fuels, a solid carbon-rich material that is usually brown or black and most often occurs in stratified sedimentary deposits.

Coal is defined as having more than 50 percent by weight (or 70 percent by volume) carbonaceous matter produced by the compaction and hardening of altered plant remains—namely, peat deposits. Different varieties of coal arise because of differences in the kinds of plant material (coal type), degree of coalification (coal rank), and range of impurities (coal grade). Although most coals occur in stratified sedimentary deposits, the deposits may later be subjected to elevated temperatures and pressures caused by igneous intrusions or deformation during orogenesis (i.e., processes of mountain building), resulting in the development of anthracite and even graphite. Although the concentration of carbon in Earth's crust does not exceed 0.1 percent by weight, it is indispensable to life and constitutes humankind's main source of energy.

Use of Coal

In Ancient Times

The discovery of the use of fire helped to distinguish humans from other animals. Early fuels were primarily wood (and charcoal derived from it), straw, and dried dung. References to the early uses of coal are meagre. Aristotle referred to "bodies which have more of earth than of smoke" and called them "coal-like substances." (It should be noted that biblical references to coal are to charcoal rather than to the rock coal.) Coal was used commercially by the Chinese long before it was used in Europe. Although no authentic record is available, coal from the Fushun mine in northeastern China may have been employed to smelt copper as early as 1000 BCE. Stones used as fuel were said to have been produced in China during the Han dynasty (206 BCE–220 CE).

In Europe

Coal cinders found among Roman ruins in England suggest that the Romans were familiar with coal use before 400 CE. The first documented proof that coal was mined in Europe was provided by the monk Reinier of Liège, who wrote (about 1200) of black earth very similar to charcoal used by metalworkers. Many references to coal mining in England and Scotland and on the European continent began to appear in the writings of the 13th century. Coal was, however, used only on a limited scale until the early 18th century, when Abraham Darby of England and others developed methods of using in blast furnaces and forges coke made from coal. Successive metallurgical and engineering developments—most notably the invention of the coal-burning steam engine by James Watt—engendered an almost insatiable demand for coal.

In the New World

Up to the time of the American Revolution, most coal used in the American colonies came from England or Nova Scotia. Wartime shortages and the needs of the munitions manufacturers, however, spurred small American coal-mining operations such as those in Virginia on the James River near Richmond. By the early 1830s mining companies had emerged along the Ohio, Illinois, and Mississippi rivers and in the Appalachian region. As in European countries, the introduction of the steam locomotive gave the American coal industry a tremendous impetus. Continued expansion of industrial activity in the United States and in Europe further promoted the use of coal.

Modern Utilization

Coal as an Energy Source

Coal is an abundant natural resource that can be used as a source of energy, as a chemical source from which numerous synthetic compounds (e.g., dyes, oils, waxes, pharmaceuticals, and pesticides) can be derived, and in the production of coke for metallurgical processes. Coal is a major source of energy in the production of electrical power using steam generation. In addition, gasification and liquefaction of coal produce gaseous and liquid fuels that can be easily transported (e.g., by pipeline) and conveniently stored in tanks. After the tremendous rise in coal use in the early 2000s, which was primarily driven by the growth of China's economy, coal use worldwide peaked in 2012. Since then coal use has experienced a steady decline, offset largely by increases in natural gas use.

Rail-mounted coal-cutting machine.

Conversion

In general, coal can be considered a hydrogen-deficient hydrocarbon with a hydrogen-to-carbon ratio near 0.8, as compared with a liquid hydrocarbons ratio near 2 (for propane, ethane, butane, and other forms of natural gas) and a gaseous hydrocarbons ratio near 4 (for gasoline). For this reason, any process used to convert coal to alternative fuels must add hydrogen (either directly or in the form of water).

Gasification refers to the conversion of coal to a mixture of gases, including carbon monoxide, hydrogen, methane, and other hydrocarbons, depending on the conditions involved. Gasification may be accomplished either in situ or in processing plants. In situ gasification is accomplished by controlled, incomplete burning of a coal bed underground while adding air and steam. The gases are withdrawn and may be burned to produce heat or generate electricity, or they may be used as synthesis gas in indirect liquefaction or the production of chemicals.

Coal liquefaction—that is, any process of turning coal into liquid products resembling crude oil— may be either direct or indirect (i.e., by using the gaseous products obtained by breaking down the chemical structure of coal). Four general methods are used for liquefaction: (1) pyrolysis and hydrocarbonization (coal is heated in the absence of air or in a stream of hydrogen), (2) solvent extraction (coal hydrocarbons are selectively dissolved and hydrogen is added to produce the desired liquids), (3) catalytic liquefaction (hydrogenation takes place in the presence of a catalyst—for example, zinc chloride), and (4) indirect liquefaction (carbon monoxide and hydrogen are combined in the presence of a catalyst).

Problems Associated with the use of Coal

Hazards of Mining and Preparation

Coal is abundant. Assuming that current rates of usage and production do not change, estimates of reserves indicate that enough coal remains to last more than 200 years. There are, however, a variety of problems associated with the use of coal.

Mining operations are hazardous. Each year hundreds of coal miners lose their lives or are seriously injured. Major mine hazards include roof falls, rock bursts, and fires and explosions. The latter result when flammable gases (such as methane) trapped in the coal are released during mining operations and accidentally are ignited. Methane may be extracted from coal beds prior to mining through the process of hydraulic fracturing (fracking), which involves high-pressure injection of fluids underground in order to open fissures in rock that would allow trapped gas or crude oil to escape into pipes that would bring the material to the surface. Methane extraction was expected to lead to safer mines and provide a source of natural gas that had long been wastedo. However, enthusiasm for this technology has been tempered with the knowledge that fracking has also been associated with groundwater contamination. In addition, miners working belowground often inhale coal dust over extended periods of time, which can result in serious health problems—for example, black lung.

Coal mines and coal-preparation plants have caused much environmental damage. Surface areas exposed during mining, as well as coal and rock waste (which were often dumped indiscriminately), weather rapidly, producing abundant sediment and soluble chemical products such as sulfuric acid and iron sulfates. Nearby streams became clogged with sediment, iron oxides stained rocks, and "acid mine drainage" caused marked reductions in the numbers of plants and animals living in the vicinity. Potentially toxic elements, leached from the exposed coal and adjacent rocks, were released into the environment. Since the 1970s, stricter laws have significantly reduced the environmental damage caused by coal mining in developed countries, though more-severe damage continues to occur in many developing countries.

Hazards of Utilization

Coal utilization can cause problems. During the incomplete burning or conversion of coal, many compounds are produced, some of which are carcinogenic. The burning of coal also produces sulfur and nitrogen oxides that react with atmospheric moisture to produce sulfuric and nitric acids—so-called acid rain. In addition, it produces particulate matter (fly ash) that can be transported by winds for many hundreds of kilometres and solids (bottom ash and slag) that must be disposed of. Trace elements originally present in the coal may escape as volatiles (e.g., chlorine and mercury) or be concentrated in the ash (e.g., arsenic and barium). Some of these pollutants can be trapped by using such devices as electrostatic precipitators, baghouses, and scrubbers. Current research on alternative means for combustion (e.g., fluidized bed combustion, magnetohydrodynamics, and low nitrogen dioxide burners) is expected to provide efficient and environmentally attractive methods for extracting energy from coal. Regardless of the means used for combustion, acceptable ways of disposing of the waste products have to be found.

The burning of all fossil fuels (oil and natural gas included) releases large quantities of carbon dioxide (CO_2) into the atmosphere. The CO_2 molecules allow the shorter-wavelength rays from the Sun to enter the atmosphere and strike Earth's surface, but they do not allow much of the long-wave radiation reradiated from the surface to escape into space. The CO_2 absorbs this upward-propagating infrared radiation and reemits a portion of it downward, causing the lower atmosphere to remain warmer than it would otherwise be. Whereas the greenhouse effect is a naturally occurring process, its enhancement due to increased release of greenhouse gases (CO_2 and other gases, such as methane and ozone) is called global warming. According to the Intergovernmental Panel on Climate Change (IPCC), there is substantial evidence that higher concentrations of CO_2 and other greenhouse gases have increased the mean temperature of Earth since 1950. This increase is probably the cause of noticeable reductions in snow cover and sea ice extent in the Northern Hemisphere. In addition, a worldwide increase in sea level and a decrease in mountain glacier extent have been documented. Technologies being considered to reduce carbon dioxide levels include biological fixation, cryogenic recovery, disposal in the oceans and aquifers, and conversion to methanol.

Coal Types and Ranks

Coals may be classified in several ways. One mode of classification is by coal type; such types have some genetic implications because they are based on the organic materials present and the coalification processes that produced the coal. The most useful and widely applied coal-classification schemes are those based on the degree to which coals have undergone coalification. Such varying degrees of coalification are generally called coal ranks (or classes). In addition to the scientific value of classification schemes of this kind, the determination of rank has a number of practical applications. Many coal properties are in part determined by rank, including the amount of heat produced during combustion, the amount of gaseous products released upon heating, and the suitability of the coals for liquefaction or for producing coke.

Macerals

Coals contain both organic and inorganic phases. The latter consist either of minerals such as quartz and clays that may have been brought in by flowing water (or wind activity) or of minerals such as pyrite and marcasite that formed in place (authigenic). Some formed in living plant tissues, and others formed later during peat formation or coalification. Some pyrite (and marcasite) is present in micrometre-sized spheroids called framboids (named for their raspberry-like shape) that formed quite early. Framboids are very difficult to remove by conventional coal-cleaning processes.

By analogy to the term mineral, British botanist Marie C. Stopes proposed in 1935 the term maceral to describe organic constituents present in coals. The corresponding ending for macerals is -inite.) Maceral nomenclature has been applied differently by some European coal petrologists who studied polished blocks of coal using reflected-light microscopy (their terminology is based on morphology, botanical affinity, and mode of occurrence) and by some North American petrologists who studied very thin slices (thin sections) of coal using transmitted-light microscopy. Various nomenclature systems have been used.

Three major maceral groups are generally recognized: vitrinite, liptinite (formerly called exinite), and inertinite. The vitrinite group is the most abundant, constituting as much as 50 to 90 percent of many North American coals. Vitrinites are derived primarily from cell walls and woody tissues. They show a wide range of reflectance values, but in individual samples these values tend to be intermediate compared with those of the other maceral groups. Several varieties are recognized—e.g., telinite (the brighter parts of vitrinite that make up cell walls) and collinite (clear vitrinite that occupies the spaces between cell walls).

The liptinite group makes up 5 to 15 percent of many coals. Liptinites are derived from waxy or resinous plant parts, such as cuticles, spores, and wound resins. Their reflectance values are usually the lowest in an individual sample. Several varieties are recognized, including sporinite (spores are typically preserved as flattened spheroids), cutinite (part of cross sections of leaves, often with crenulated surfaces), and resinite (ovoid and sometimes translucent masses of resin). The liptinites may fluoresce (i.e., luminesce because of absorption of radiation) under ultraviolet light, but with increasing rank their optical properties approach those of the vitrinites, and the two groups become indistinguishable.

The inertinite group makes up 5 to 40 percent of most coals. Their reflectance values are usually the highest in a given sample. The most common inertinite maceral is fusinite, which has a charcoal-like appearance with obvious cell texture. The cells may be either empty or filled with mineral matter, and the cell walls may have been crushed during compaction (bogen texture). Inertinites are derived from strongly altered or degraded plant material that is thought to have been produced during the formation of peat; in particular, charcoal produced by a fire in a peat swamp is preserved as fusinite.

Coal Rock Types

Coals may be classified on the basis of their macroscopic appearance (generally referred to as coal rock type, lithotype, or kohlentype). Four main types are recognized:

1. Vitrain, which is characterized by a brilliant black lustre and composed primarily of the maceral group vitrinite, which is derived from the woody tissue of large plants. Vitrain is brittle and tends to break into angular fragments; however, thick vitrain layers show conchoidal fractures (that is, curving fractures that resemble the interior of a seashell) when broken. Vitrain occurs in narrow, sometimes markedly uniform, bright bands that are about 3 to 10 mm (about 0.1 to 0.4 inch) thick. Vitrain probably formed under somewhat drier surface conditions than did the lithotypes clarain and durain. On burial, stagnant groundwater prevented the complete decomposition of the woody plant tissues.

2. Clarain, which has an appearance between those of vitrain and durain and is characterized by alternating bright and dull black laminae (thin layers, each commonly less than 1 mm thick). The brightest layers are composed chiefly of the maceral vitrinite and the duller layers of the other maceral groups, liptinite and inertinite. Clarain exhibits a silky lustre less brilliant than that of vitrain. It seems to have originated under conditions that alternated between those in which durain and vitrain formed.

3. Durain, which is characterized by a hard granular texture and composed of the maceral groups liptinite and inertinite as well as relatively large amounts of inorganic minerals. Durain occurs in layers more than 3 to 10 mm (about 0.1 to 0.4 inch) thick, although layers more than 10 cm (about 4 inches) thick have been recognized. Durains are usually dull black to dark gray in colour. Durain is thought to have formed in peat deposits below water level, where only liptinite and inertinite components resisted decomposition and where inorganic minerals accumulated from sedimentation.

4. Fusain, which is commonly found in silky and fibrous lenses that are only millimetres thick and centimetres long. Most fusain is extremely soft and crumbles readily into a fine, sootlike powder that soils the hands. Fusain is composed mainly of fusinite (carbonized woody plant tissue) and semifusinite from the maceral group inertinite, which is rich in carbon and highly reflective. It closely resembles charcoal, both chemically and physically, and is believed to have been formed in peat deposits swept by forest fires, by fungal activity that generated intense heat, or by subsurface oxidation of coal.

Banded and Nonbanded Coals

The term coal type is employed to distinguish between banded coals and nonbanded coals. Banded coals contain varying amounts of vitrinite and opaque material. They are made up of less than 5 percent anthraxylon (the translucent glossy jet-black material in bituminous coal) that alternates with thin bands of dull coal called attritus. Banded coals include bright coal, which contains more than 80 percent vitrinite, and splint coal, which contains more than 30 percent opaque matter. The nonbanded varieties include boghead coal, which has a high percentage of algal remains, and cannel coal, which has a high percentage of spores in its attritus (that is, pulverized or finely divided matter). The anthraxylon content in nonbanded coals exceeds 5 percent. The usage of all the above terms is quite subjective.

Ranking by Coalification

Hydrocarbon Content

The oldest coal-classification system was based on criteria of chemical composition. Developed in 1837 by the French chemist Henri-Victor Regnault, it was improved in later systems that classified coals on the basis of their hydrogen and carbon content. However, because the relationships between chemistry and other coal properties are complex, such classifications are rarely used for practical purposes today.

Chemical Content and Properties

Coal is divided into a number of ranks to help buyers such as electrical utilities assess the calorific value and volatile matter content of each unit of coal they purchase. The most commonly employed systems of classification are those based on analyses that can be performed relatively easily in the laboratory—for example, determining the percentage of volatile matter lost upon heating to about 950 °C (about 1,750 °F) or the amount of heat released during combustion of the coal under standard conditions. ASTM International (formerly the American Society for Testing and Materials) assigns ranks to coals on the basis of fixed carbon content, volatile matter content, and calorific value. In addition to the major ranks (lignite, subbituminous, bituminous, and anthracite), each rank may be divided into coal groups such as high-volatile A bituminous coal. These categories differ slightly between countries; however, the ranks are often comparable with respect to moisture, volatile matter content, and heating value. Other designations, such as coking coal and steam coal, have been applied to coals, and they also tend to differ from country to country.

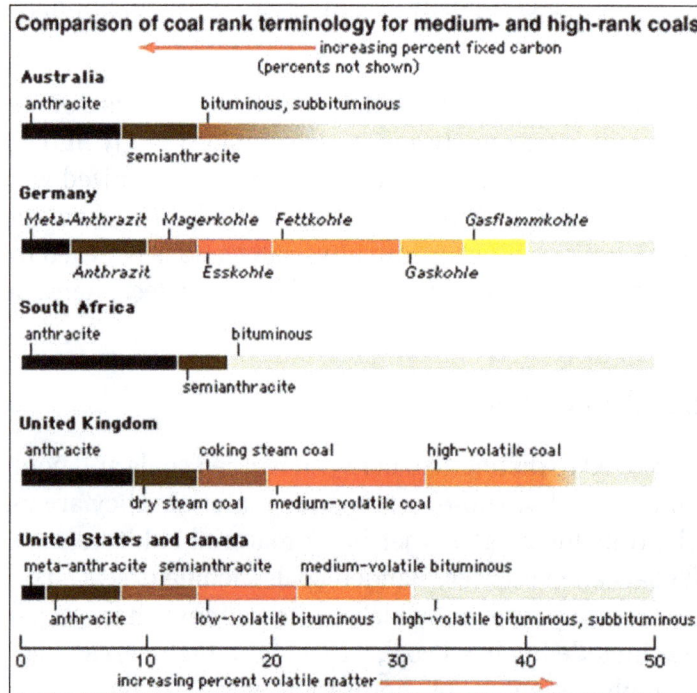

Comparison of coal-rank terminologies by country.

Virtually all classification systems use the percentage of volatile matter present to distinguish coal ranks. In the ASTM classification, high-volatile A bituminous (and higher ranks) are classified on the basis of their volatile matter content. Coals of lower rank are classified primarily on the basis of their heat values, because of their wide ranges in volatile matter content (including moisture). The agglomerating character of a coal refers to its ability to soften and swell when heated and to form cokelike masses that are used in the manufacture of steel. The most suitable coals for agglomerating purposes are in the bituminous rank.

Coal analyses may be presented in the form of "proximate" and "ultimate" analyses, whose analytical conditions are prescribed by organizations such as ASTM. A typical proximate analysis includes the moisture, ash, volatile matter, and fixed carbon contents. (Fixed carbon is the material, other than ash, that does not vaporize when heated in the absence of air. It is usually determined by subtracting the sum of the first three values—moisture, ash, and volatile matter—in weight percent from 100 percent.) It is important for economic reasons to know the moisture and ash contents of a coal because they do not contribute to the heating value of a coal. In most cases ash becomes an undesirable residue and a source of pollution, but for some purposes (e.g., use as a chemical source or for coal liquefaction) the presence of mineral matter may be desirable. Most of the heat value of a coal comes from its volatile matter, excluding moisture, and fixed carbon content. For most coals it is necessary to measure the actual amount of heat released upon combustion (expressed in megajoules per kilogram or British thermal units per pound).

Ultimate analyses are used to determine the carbon, hydrogen, sulfur, nitrogen, ash, oxygen, and moisture contents of a coal. For specific applications, other chemical analyses may be employed. These may involve, for example, identifying the forms of sulfur present. Sulfur may occur in the form of sulfide minerals (pyrite and marcasite), sulfate minerals (gypsum), or organically bound sulfur. In other cases the analyses may involve determining the trace elements present (e.g., mercury, chlorine), which may influence the suitability of a coal for a particular purpose or help to establish methods for reducing environmental pollution and so forth.

Coal-forming Materials

Plant Matter

It is generally accepted that most coals formed from plants that grew in and adjacent to swamps in warm, humid regions. Material derived from these plants accumulated in low-lying areas that remained wet most of the time and was converted to peat through the activity of microorganisms. (It should be noted that peat can occur in temperate regions [e.g., Ireland and the state of Michigan in the United States] and even in subarctic regions [e.g., the Scandinavian countries].) Under certain conditions this organic material continued to accumulate and was later converted into coal. Much of the plant matter that accumulates on the surface of Earth is never converted to peat or to coal, because it is removed by fire or organic decomposition. Hence, the vast coal deposits found in ancient rocks must represent periods during which several favourable biological and physical processes occurred at the same time.

Evidence that coal was derived from plants comes from three principal sources. First, lignites, the lowest coal rank, often contain recognizable plant remains. Second, sedimentary rock layers above, below, and adjacent to coal seams contain plant fossils in the form of impressions and carbonized films (e.g., leaves and stems) and casts of larger parts such as roots, branches, and trunks. Third, even coals of advanced rank may reveal the presence of precursor plant material. When examined microscopically in thin sections or polished blocks, cell walls, cuticles (the outer wall of leaves), spores, and other structures can still be recognized. Algal and fungal remains also may be present. (Algae are major components in boghead coal, a type of sapropelic coal.)

Fossil Record

Anthracite (the highest coal rank) material, which appears to have been derived from algae, is known from the Proterozoic Eon (approximately 2.5 billion to 541 million years ago) of Precambrian time. Siliceous rocks of the same age contain fossil algae and fungi. These early plants were primarily protists (solitary or aggregate unicellular organisms that include yellow-green algae, golden-brown algae, and diatoms) that lived in aqueous environments. By the Silurian Period (443.8 million to 419.2 million years ago), plants had developed the ability to survive on land and had invaded the planet's coastal areas.

Evidence for coastal forests is preserved in strata of the Ordovician Period (485.4 million to 443.8 million years ago). By the latter half of the Paleozoic Era, plants had undergone extensive evolution and occupied many previously vacant environments (this phenomenon is sometimes called adaptive radiation).

There were two major eras of coal formation in geologic history. The older includes the Carboniferous Period (extending from 358.9 million to 298.9 million years ago and often divided into the Mississippian and Pennsylvanian subperiods) and the Permian Period (from approximately 298.9 million to 251.9 million years ago) of the Paleozoic Era. Much of the bituminous coal of eastern North America and Europe is Carboniferous in age. Most coals in Siberia, eastern Asia, and Australia are of Permian origin.

Pennsylvanian coal forest diorama.

The younger era of coal formation began about 135 million years ago during the Cretaceous Period and reached its peak approximately 66 million to 2.6 million years ago, during the Paleogene and Neogene periods of the Cenozoic Era. Most of the coals that formed during this later era are lignites and subbituminous (brown) coals. These are widespread in western North America (including Alaska), southern France and central Europe, Japan, and Indonesia.

Late Paleozoic flora included sphenopsids, lycopsids, pteropsids, and the Cordaitales. The sphenopsid Calamites grew as trees in swamps. Calamites had long, jointed stems with sparse foliage. The lycopsids included species of Lepidodendron and Sigillaria (up to 30 metres [about 100 feet] tall) that grew in somewhat drier areas. Pteropsids included both true ferns (Filicineae) and extinct seed ferns (Pteridospermaphyta), which grew in relatively dry environments. The Cordaitales, which had tall stems and long, narrow, palmlike leaves, also favoured drier areas. During the Cretaceous and Cenozoic the angiosperms (flowering plants) evolved, producing a diversified flora from which the younger coals developed.

Formation Processes

Peat

Although peat is used as a source of energy, it is not usually considered a coal. It is the precursor material from which coals are derived, and the process by which peat is formed is studied in existing swamps in many parts of the world (e.g., in the Okefenokee Swamp of Georgia, U.S. and along the southwestern coast of New Guinea). The formation of peat is controlled by several factors, including: (1) the evolutionary development of plant life, (2) the climatic conditions (warm enough to sustain plant growth and wet enough to permit the partial decomposition of the plant material and preserve the peat), and (3) the physical conditions of the area (its geographic position relative to the sea or other bodies of water, rates of subsidence or uplift, and so forth). Warm moist climates are thought to produce broad bands of bright coal, a type of bituminous coal characterized by its fine banding and high concentrations of nitrogen, sulfur, and moisture. Cooler temperate climates, on the other hand, are thought to produce detrital coal (which is thought to be the remains of preexisting coal beds) with relatively little bright coal.

Inundated peat bog.

Initially, the area on which a future coal seam may be developed must be uplifted so that plant growth can be established. Areas near seacoasts or low-lying areas near streams stay moist enough for peat to form, but elevated swamps (some bogs and moors) can produce peat only if the annual precipitation exceeds annual evaporation and little percolation or drainage occurs. Thick peat deposits necessary for coal formation develop at sites where the following conditions exist: slow, continuous subsidence; the presence of such natural structures as levees, beaches, and bars that give protection from frequent inundation; and a restricted supply of incoming sediments that would interrupt peat formation. In such areas the water may become quite stagnant (except for a few rivers traversing the swamp), and plant material can continue to accumulate. Microorganisms attack the plant material and convert it to peat. Very close to the surface where oxygen is still readily available (aerobic, or oxidizing, conditions), the decomposition of the plant material produces mostly gaseous and liquid products. With increasing depth, however, the conditions become increasingly anaerobic (reducing), and molds and peats develop. The process of peat formation—biochemical coalification—is most active in the upper few metres of a peat deposit. Fungi are not found below about 0.5 metre (about 18 inches), and most forms of microbial life are eliminated at depths below about 10 metres (about 30 feet). If either the rate of subsidence or the rate of influx of new sediment increases, the peat will be buried and soon thereafter the coalification process—geochemical coalification—begins. The cycle may be repeated many times, which accounts for the numerous coal seams found in some sedimentary basins.

Coalification

The general sequence of coalification is from lignite to subbituminous to bituminous to anthracite. Since microbial activity ceases within a few metres of Earth's surface, the coalification process must be controlled primarily by changes in physical conditions that take place with depth. Some coal characteristics are determined by events that occur during peat formation—e.g., charcoal-like material in coal is attributed to fires that occurred during dry periods while peat was still forming.

Brown-coal (lignite) pit in Eschweiler in the Rhenish field between Cologne and Aachen.

Three major physical factors—duration, increasing temperature, and increasing pressure—may influence the coalification process. In laboratory experiments artificially prepared coals are influenced by the duration of the experiment, but in nature the length of time is substantially longer and the overall effect of time remains undetermined. Low-rank coal (i.e., brown coal) in the Moscow Basin was deposited during Carboniferous time but was not buried deeply and never reached a higher rank. The most widely accepted explanation is that coalification takes place in response to increasing temperature. In general, temperature increases with depth. This geothermal gradient averages about 30 °C (about 85 °F) per kilometre, but the gradient ranges from less than 10 °C (50 °F) per kilometre in regions undergoing very rapid subsidence to more than 100 °C (212 °F) per kilometre in areas of igneous activity. Measurements of thicknesses of sedimentary cover and corresponding coal ranks suggest that temperatures lower than 200 °C (about 390 °F) are sufficient to produce coal of anthracite rank. The effect of increasing pressure due to depth of burial is not considered to cause coalification. In fact, increasing overburden pressure might have the opposite effect if volatile compounds such as methane that must escape during coalification are retained. Pressure may influence the porosity and moisture content of coal.

Structure and Properties of Coal

Organic Compounds

The plant material from which coal is derived is composed of a complex mixture of organic compounds, including cellulose, lignin, fats, waxes, and tannins. As peat formation and coalification proceed, these compounds, which have more or less open structures, are broken down, and new compounds—primarily aromatic (benzenelike) and hydroaromatic—are produced. In vitrinite these compounds are connected by cross-linking oxygen, sulfur, and molecules such as methylene. During coalification, volatile phases rich in hydrogen and oxygen (e.g., water, carbon dioxide, and methane) are produced and escape from the mass; hence, the coal becomes progressively richer in carbon. The classification of coal by rank is based on these changes—i.e., as coalification proceeds, the amount of volatile matter gradually decreases and the amount of fixed carbon increases. As volatiles are expelled, more carbon-to-carbon linkages occur in the remaining coal until, having reached the anthracite rank, it takes on many of the characteristics of the end product of the metamorphism of carbonaceous material—namely, graphite. Coals pass through several structural states as the bonds between the aromatic nuclei increase.

Properties

Many of the properties of coal are strongly rank-dependent, although other factors such as maceral composition and the presence of mineral matter also influence its properties. Several techniques have been developed for studying the physical and chemical properties of coal, including density measurements, X-ray diffraction, scanning and transmission electron microscopy, infrared spectrophotometry, mass spectroscopy, gas chromatography, thermal analysis, and electrical, optical, and magnetic measurements.

Density

Knowledge of the physical properties of coal is important in coal preparation and utilization. For example, coal density ranges from approximately 1.1 to about 1.5 megagrams per cubic metre, or grams per cubic centimetre (1 megagram per cubic metre equals 1 gram per cubic centimetre). Coal is slightly denser than water (1.0 megagram per cubic metre) and significantly less dense than most rock and mineral matter (e.g., shale has a density of about 2.7 megagrams per cubic metre and pyrite of 5.0 megagrams per cubic metre). Density differences make it possible to improve the quality of a coal by removing most of the rock matter and sulfide-rich fragments by means of heavy liquid separation (fragments with densities greater than about 1.5 megagrams per cubic metre settle out while the coal floats on top of the liquid). Devices such as cyclones and shaker tables also separate coal particles from rock and pyrite on the basis of their different densities.

Porosity

Coal density is controlled in part by the presence of pores that persist throughout coalification. Measurement of pore sizes and pore distribution is difficult; however, there appear to be three size ranges of pores: (1) macropores (diameter greater than 50 nanometres), (2) mesopores (diameter 2 to 50 nanometres), and (3) micropores (diameter less than 2 nanometres). (One nanometre is equal to $10-9$ metre.) Most of the effective surface area of a coal—about 200 square metres per gram—is not on the outer surface of a piece of coal but is located inside the coal in its pores. The presence of pore space is important in the production of coke, gasification, liquefaction, and the generation of high-surface-area carbon for purifying water and gases. From the standpoint of safety, coal pores may contain significant amounts of adsorbed methane that may be released during mining operations and form explosive mixtures with air. The risk of explosion can be reduced by adequate ventilation during mining or by prior removal of coal-bed methane.

Reflectivity

An important property of coal is its reflectivity (or reflectance)—i.e., its ability to reflect light. Reflectivity is measured by shining a beam of monochromatic light (with a wavelength of 546 nanometres) on a polished surface of the vitrinite macerals in a coal sample and measuring the percentage of the light reflected with a photometer. Vitrinite is used because its reflectivity changes gradually with increasing rank. Fusinite reflectivities are too high due to its origin as charcoal, and liptinites tend to disappear with increasing rank. Although little of the incident light is reflected (ranging from a few tenths of a percent to 12 percent), the value increases with rank and can be used to determine the rank of most coals without measuring the percentage of volatile matter present.

The study of coals (and coaly particles called phyterals) in sedimentary basins containing oil and/ or gas reveals a close relationship between coalification and the maturation of liquid and gaseous hydrocarbons. During the initial stages of coalification (to a reflectivity of almost 0.5 and near the boundary between subbituminous and high-volatile C bituminous coal), hydrocarbon generation produces chiefly methane. The maximum generation of liquid petroleum occurs during the development of high-volatile bituminous coals (in the reflectivity range from roughly 0.5 to about 1.3). With increasing depth and temperature, petroleum liquids break down and, finally, only natural gas (methane) remains. Geologists can use coal reflectivity to anticipate the potential for finding liquid or gaseous hydrocarbons as they explore for petroleum.

Other Properties

Other properties, such as hardness, grindability, ash-fusion temperature, and free-swelling index (a visual measurement of the amount of swelling that occurs when a coal sample is heated in a covered crucible), may affect coal mining and preparation, as well as the way in which a coal is used. Hardness and grindability determine the kinds of equipment used for mining, crushing, and grinding coals in addition to the amount of power consumed in their operation. Ash-fusion temperature influences furnace design and operating conditions. The free-swelling index provides preliminary information concerning the suitability of a coal for coke production.

World Distribution of Coal

General Occurrence

Coal is a widespread resource of energy and chemicals. Although terrestrial plants necessary for the development of coal did not become abundant until Carboniferous time (358.9 million to 298.9 million years ago), large sedimentary basins containing rocks of Carboniferous age and younger are known on virtually every continent, including Antarctica (not shown on the map). The presence of large coal deposits in regions that now have arctic or subarctic climates (such as Alaska and Siberia) is due to climatic changes and to the tectonic motion of crustal plates that moved ancient continental masses over Earth's surface, sometimes through subtropical and even tropical regions. Coal is absent in some areas (such as Greenland and much of northern Canada) because the rocks found there predate the Carboniferous Period and these regions, known as continental shields, lacked the abundant terrestrial plant life needed for the formation of major coal deposits.

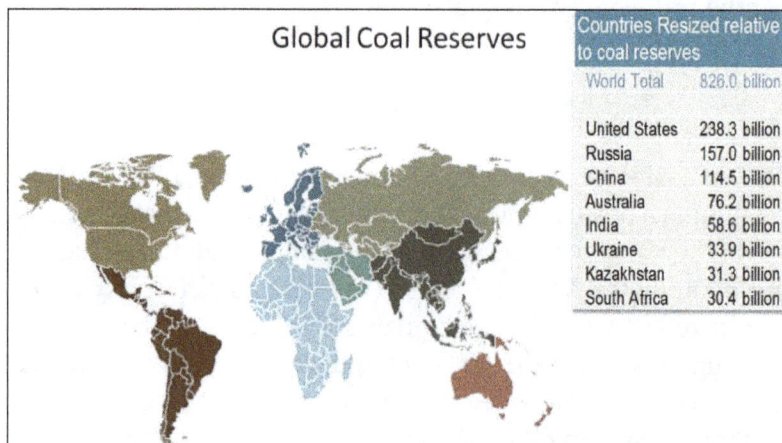

Global Coal Reserves

Countries Resized relative to coal reserves	
World Total	826.0 billion
United States	238.3 billion
Russia	157.0 billion
China	114.5 billion
Australia	76.2 billion
India	58.6 billion
Ukraine	33.9 billion
Kazakhstan	31.3 billion
South Africa	30.4 billion

Resources and Reserves

World coal reserves and resources are difficult to assess. Although some of the difficulty stems from the lack of accurate data for individual countries, two fundamental problems make these estimates difficult and subjective. The first problem concerns differences in the definition of terms such as proven reserves (generally only those quantities that are recoverable) and geological resources (generally the total amount of coal present, whether or not recoverable at present).

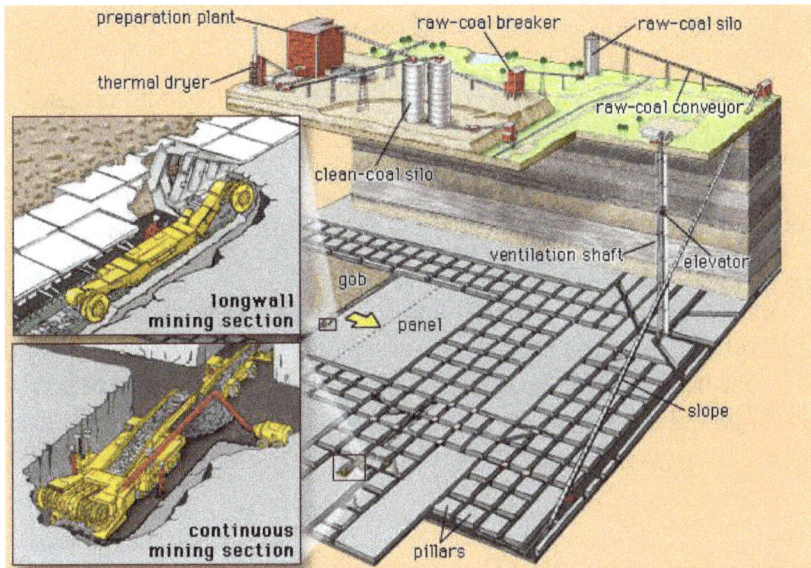

Schematic diagram of an underground coal mine, showing surface facilities, access shaft, and the room-and-pillar and longwall mining methods.

The proven reserves for any commodity should provide a reasonably accurate estimate of the amount that can be recovered under existing operating and economic conditions. To be economically mineable, a coal bed must have a minimum thickness (about 0.6 metre; 2 feet) and be buried less than some maximum depth (roughly 2,000 metres; 6,600 feet) below Earth's surface. These values of thickness and depth are not fixed but change with coal quality, demand, the ease with which overlying rocks can be removed (in surface mining) or a shaft sunk to reach the coal seam (in underground mining), and so forth. The development of new mining techniques may increase the amount of coal that can be extracted relative to the amount that cannot be removed. For example, in underground mining (which accounts for about 60 percent of world coal production), conventional mining methods leave behind large pillars of coal to support the overlying rocks and recover only about half of the coal present. On the other hand, longwall mining, in which the equipment removes continuous parallel bands of coal, may recover nearly all the coal present.

The second problem, which concerns the estimation of reserves, is the rate at which a commodity is consumed. When considering the worldwide reserves of coal, the number of years that coal will be available may be more important than the total amount of coal resources. At present rates of consumption, world coal reserves should last more than 300–500 years. A large amount of additional coal is present in Earth but cannot be recovered at this time. These resources, sometimes called "geologic resources," are even more difficult to estimate, but they are thought to be as much as 15 times greater than the amount of proven reserves.

Petroleum

Petroleum is a complex mixture of hydrocarbons that occur in Earth in liquid, gaseous, or solid form. The term is often restricted to the liquid form, commonly called crude oil, but, as a technical term, petroleum also includes natural gas and the viscous or solid form known as bitumen, which is found in tar sands. The liquid and gaseous phases of petroleum constitute the most important of the primary fossil fuels.

Liquid and gaseous hydrocarbons are so intimately associated in nature that it has become customary to shorten the expression "petroleum and natural gas" to "petroleum" when referring to both. The word petroleum was first used in 1556 in a treatise published by the German mineralogist Georg Bauer, known as Georgius Agricola.

The burning of all fossil fuels (coal and biomass included) releases large quantities of carbon dioxide (CO_2) into the atmosphere. The CO_2 molecules do not allow much of the long-wave solar radiation absorbed by Earth's surface to reradiate from the surface and escape into space. The CO_2 absorbs upward-propagating infrared radiation and reemits a portion of it downward, causing the lower atmosphere to remain warmer than it would otherwise be. This phenomenon has the effect of enhancing Earth's natural greenhouse effect, producing what scientists refer to as anthropogenic (human-generated) global warming. There is substantial evidence that higher concentrations of CO_2 and other greenhouse gases have contributed greatly to the increase of Earth's near-surface mean temperature since 1950.

Uses

Exploitation of Surface Seeps

Small surface occurrences of petroleum in the form of natural gas and oil seeps have been known from early times. The ancient Sumerians, Assyrians, and Babylonians used crude oil, bitumen, and asphalt ("pitch") collected from large seeps at Tuttul (modern-day Hīt) on the Euphrates for many purposes more than 5,000 years ago. Liquid oil was first used as a medicine by the ancient Egyptians, presumably as a wound dressing, liniment, and laxative. The Assyrians used bitumen as a means of punishment by pouring it over the heads of lawbreakers.

Oil products were valued as weapons of war in the ancient world. The Persians used incendiary arrows wrapped in oil-soaked fibres at the siege of Athens in 480 BCE. Early in the Common Era the Arabs and Persians distilled crude oil to obtain flammable products for military purposes. Probably as a result of the Arab invasion of Spain, the industrial art of distillation into illuminants became available in western Europe by the 12th century.

Several centuries later, Spanish explorers discovered oil seeps in present-day Cuba, Mexico, Bolivia, and Peru. Oil seeps were plentiful in North America and were also noted by early explorers in what are now New York and Pennsylvania, where American Indians were reported to have used the oil for medicinal purposes.

Extraction from Underground Reservoirs

Until the beginning of the 19th century, illumination in the United States and in many other countries was little improved over that which was known during the times of the Mesopotamians, Greeks, and Romans. Greek and Roman lamps and light sources often relied on the oils produced by animals (such as fish and birds) and plants (such as olive, sesame, and nuts). Timber was also ignited to produce illumination. Since timber was scarce in Mesopotamia, "rock asphalt" (sandstone or limestone infused with bitumen or petroleum residue) was mined and combined with sand and fibres for use in supplementing building materials. The need for better illumination that accompanied the increasing development of urban centres made it necessary to search for new sources of oil, especially since whales, which had long provided fuel for lamps, were becoming harder and harder to find. By the mid-19th century kerosene, or coal oil, derived from coal was in common use in both North America and Europe.

The Industrial Revolution brought an ever-growing demand for a cheaper and more convenient source of lubricants as well as of illuminating oil. It also required better sources of energy. Energy had previously been provided by human and animal muscle and later by the combustion of such solid fuels as wood, peat, and coal. These were collected with considerable effort and laboriously transported to the site where the energy source was needed. Liquid petroleum, on the other hand, was a more easily transportable source of energy. Oil was a much more concentrated and flexible form of fuel than anything previously available.

The stage was set for the first well specifically drilled for oil, a project undertaken by American entrepreneur Edwin L. Drake in northwestern Pennsylvania. The completion of the well in August 1859 established the groundwork for the petroleum industry and ushered in the closely associated modern industrial age. Within a short time, inexpensive oil from underground reservoirs was being processed at already existing coal oil refineries, and by the end of the century oil fields had been discovered in 14 states from New York to California and from Wyoming to Texas. During the same period, oil fields were found in Europe and East Asia as well.

Significance of Petroleum in Modern Times

At the beginning of the 20th century, the Industrial Revolution had progressed to the extent that the use of refined oil for illuminants ceased to be of primary importance. The oil and gas industry became the major supplier of energy largely because of the advent of the internal-combustion engine, especially those in automobiles. Although oil constitutes a major petrochemical feedstock, its primary importance is as an energy source on which the world economy depends.

The significance of oil as a world energy source is difficult to overdramatize. The growth in energy production during the 20th century was unprecedented, and increasing oil production has been by far the major contributor to that growth. By the 21st century an immense and intricate value chain was moving approximately 100 million barrels of oil per day from producers to consumers. The production and consumption of oil is of vital importance to international relations and has frequently been a decisive factor in the determination of foreign policy. The position of a country in this system depends on its production capacity as related to its consumption. The possession of oil deposits is sometimes the determining factor between a rich and a poor country. For any country, the presence or absence of oil has major economic consequences.

On a timescale within the span of prospective human history, the utilization of oil as a major source of energy will be a transitory affair lasting only a few centuries. Nonetheless, it will have been an affair of profound importance to world industrialization.

Properties of Hydrocarbons

Hydrocarbon Content

Although oil consists basically of compounds of only two elements, carbon and hydrogen, these elements form a large variety of complex molecular structures. Regardless of physical or chemical variations, however, almost all crude oil ranges from 82 to 87 percent carbon by weight and 12 to 15 percent hydrogen. The more-viscous bitumens generally vary from 80 to 85 percent carbon and from 8 to 11 percent hydrogen.

Crude oil is an organic compound divided primarily into alkenes with single-bond hydrocarbons of the form C_nH_{2n+2} or aromatics having six-ring carbon-hydrogen bonds, C_6H_6. Most crude oils are grouped into mixtures of various and seemingly endless proportions. No two crude oils from different sources are completely identical.

The alkane paraffinic series of hydrocarbons, also called the methane (CH_4) series, comprises the most common hydrocarbons in crude oil. The major constituents of gasoline are the paraffins that are liquid at normal temperatures but boil between 40 °C and 200 °C (100 °F and 400 °F). The residues obtained by refining lower-density paraffins are both plastic and solid paraffin waxes.

The naphthenic series has the general formula $CnH2n$ and is a saturated closed-ring series. This series is an important part of all liquid refinery products, but it also forms most of the complex residues from the higher boiling-point ranges. For this reason, the series is generally heavier. The residue of the refining process is an asphalt, and the crude oils in which this series predominates are called asphalt-base crudes.

The aromatic series is an unsaturated closed-ring series. Its most common member, benzene (C_6H_6), is present in all crude oils, but the aromatics as a series generally constitute only a small percentage of most crudes.

Nonhydrocarbon Content

In addition to the practically infinite mixtures of hydrocarbon compounds that form crude oil, sulfur, nitrogen, and oxygen are usually present in small but often important quantities. Sulfur is the third most abundant atomic constituent of crude oils. It is present in the medium and heavy fractions of crude oils. In the low and medium molecular ranges, sulfur is associated only with carbon and hydrogen, while in the heavier fractions it is frequently incorporated in the large polycyclic molecules that also contain nitrogen and oxygen. The total sulfur in crude oil varies from below 0.05 percent (by weight), as in some Venezuelan oils, to about 2 percent for average Middle Eastern crudes and up to 5 percent or more in heavy Mexican or Mississippi oils. Generally, the higher the specific gravity of the crude oil (which determines whether crude is heavy, medium, or light), the greater its sulfur content. The excess sulfur is removed from crude oil prior to refining, because sulfur oxides released into the atmosphere during the combustion of oil would constitute a major pollutant, and they also act as a significant corrosive agent in and on oil processing equipment.

The oxygen content of crude oil is usually less than 2 percent by weight and is present as part of the heavier hydrocarbon compounds in most cases. For this reason, the heavier oils contain the most oxygen. Nitrogen is present in almost all crude oils, usually in quantities of less than 0.1 percent by weight. Sodium chloride also occurs in most crudes and is usually removed like sulfur.

Many metallic elements are found in crude oils, including most of those that occur in seawater. This is probably due to the close association between seawater and the organic forms from which oil is generated. Among the most common metallic elements in oil are vanadium and nickel, which apparently occur in organic combinations as they do in living plants and animals.

Crude oil also may contain a small amount of decay-resistant organic remains, such as siliceous skeletal fragments, wood, spores, resins, coal, and various other remnants of former life.

Physical Properties

Crude oil consists of a closely related series of complex hydrocarbon compounds that range from gasoline to heavy solids. The various mixtures that constitute crude oil can be separated by distillation under increasing temperatures into such components as (from light to heavy) gasoline, kerosene, gas oil, lubricating oil, residual fuel oil, bitumen, and paraffin.

Crude oils vary greatly in their chemical composition. Because they consist of mixtures of thousands of hydrocarbon compounds, their physical properties—such as specific gravity, colour, and viscosity (resistance of a fluid to a change in shape)—also vary widely.

Specific Gravity

Crude oil is immiscible with and lighter than water; hence, it floats. Crude oils are generally classified as bitumens, heavy oils, and medium and light oils on the basis of specific gravity (i.e., the ratio of the weight of equal volumes of the oil and pure water at standard conditions, with pure water considered to equal 1) and relative mobility. Bitumen is an immobile degraded remnant of ancient petroleum; it is present in oil sands and does not flow into a well bore. Heavy crude oils have enough mobility that, given time, they can be obtained through a well bore in response to enhanced recovery methods—that is, techniques that involve heat, gas, or chemicals that lower the viscosity of petroleum or drive it toward the production well bore. The more-mobile medium and light oils are recoverable through production wells.

The widely used American Petroleum Institute (API) gravity scale is based on pure water, with an arbitrarily assigned API gravity of 10°. (API gravities are unitless and are often referred to in degrees; they are calculated by multiplying the inverse of the specific gravity of a liquid at 15.5 °C [60 °F] by 141.5.) Liquids lighter than water, such as oil, have API gravities numerically greater than 10°. Crude oils below 22.3° API gravity are usually considered heavy, whereas the conventional crudes with API gravities between 22.3° and 31.1° are regarded as medium, and light oils have an API gravity above 31.1°. Optimum refinery crude oils considered the best are 40° to 45°, since anything lighter is composed of lower carbon numbers (the number of carbon atoms per molecule of material). Refinery crudes heavier than 35° API have higher carbon numbers and are more complicated to break down or process for optimal octane gasolines and diesel fuels. Early 21st-century production trends showed, however, a shift in emphasis toward heavier crudes as conventional oil reserves (that is, those not produced from source rock) declined and a greater volume of heavier oils was developed.

Boiling and Freezing Points

Because oil is always at a temperature above the boiling point of some of its compounds, the more volatile constituents constantly escape into the atmosphere unless confined. It is impossible to refer to a common boiling point for crude oil because of the widely differing boiling points of its numerous compounds, some of which may boil at temperatures too high to be measured.

By the same token, it is impossible to refer to a common freezing point for crude oil because the individual compounds solidify at different temperatures. However, the pour point—the temperature below which crude oil becomes plastic and will not flow—is important to recovery and transport and is always determined. Pour points range from 32 °C to below −57 °C (90 °F to below −70 °F).

Measurement Systems

In the United States, crude oil is measured in barrels of 42 gallons each; the weight per barrel of API 30° light oil is about 306 pounds. In many other countries, crude oil is measured in metric tons. For crude oil having the same gravity, a metric ton is equal to approximately 252 imperial gallons or about 7.2 U.S. barrels.

Resources and Reserves

Reservoirs formed by traps or seeps contain hydrocarbons that are further defined as either resources or reserves. Resources are the total amount of all possible hydrocarbons estimated from formations before wells are drilled. In contrast, reserves are subsets of resources; the sizes of reserves are determined by how economically or technologically feasible they are to extract petroleum from and use under current technological and economic conditions. Reserves are classified into various categories based on the amount that is likely to be extracted. Proven reserves have the highest certainty of successful extraction for commercial use (more than 90 percent), whereas successful extraction regarding probable and possible reserves for commercial use are estimated at 50 percent and between 10 and 50 percent respectively.

The broader category of resources includes both conventional and unconventional petroleum plays (or accumulations) as identified by analogs—that is, fields or reservoirs where there are few or no wells drilled but which are similar geologically to producing fields. For resources where some exploration or discovery activity has taken place, estimates of the size and number of undiscovered hydrocarbon accumulations are determined by technical experts and geoscientists as well as from measurements derived from geologic framework modeling and visualizations.

Unconventional Oil

Within the vast unconventional resources category, there are several different types of hydrocarbons, including very heavy oils, oil sands, oil shales, and tight oils. By the early 21st century, technological advances had created opportunities to convert what were once undeveloped resource plays into economic reserves.

Very heavy crudes have become economical. Those having less than 15° API can be extracted by working with natural reservoir temperatures and pressures, provided that the temperatures and pressures are high enough. Such conditions occur in Venezuela's Orinoco basin, for example. On the other hand, other very heavy crudes, such as certain Canadian crude oils, require the injection of steam from horizontal wells that also allow for gravity drainage and recovery.

Tar sands differ from very heavy crude oil in that bitumen adheres to sand particles with water. In order to convert this resource into a reserve, surface mining or subsurface steam injection into the reservoir must take place first. Later the extracted material is processed at an extraction plant capable of separating the oil from the sand, fines (very small particles), and water slurry.

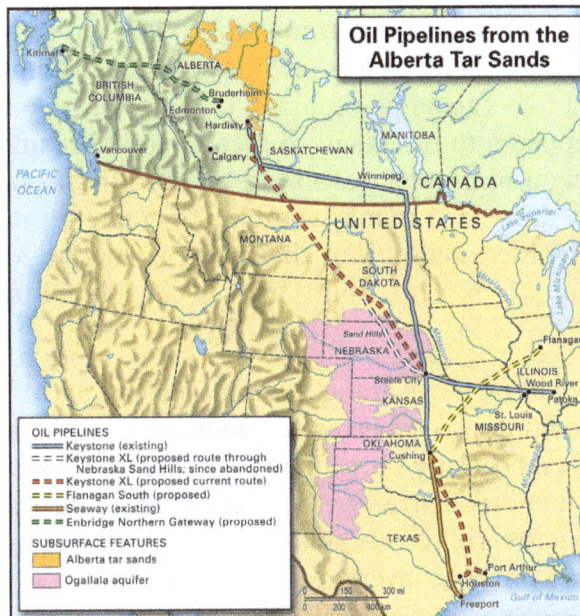

Alberta tar sands: The location of the Alberta tar sands region and its associated oil pipelines.

Oil shales make up an often misunderstood category of unconventional oils in that they are often confused with coal. Oil shale is an inorganic, nonporous rock containing some organic kerogen. While oil shales are similar to the source rock producing petroleum, they are different in that they contain up to 70 percent kerogen. In contrast, source rock tight oils contain only about 1 percent kerogen. Another key difference between oil shales and the tight oil produced from source rock is that oil shale is not exposed to sufficiently high temperatures to convert the kerogen to oil. In this sense, oil shales are hybrids of source rock oil and coal. Some oil shales can be burned as a solid. However, they are sooty and possess an extremely high volatile matter content when burned. Thus, oil shales are not used as solid fuels, but, after they are strip-mined and distilled, they are used as liquid fuels. Compared with other unconventional oils, oil shale cannot be extracted practically through hydraulic fracturing or thermal methods at present.

Shale oil is a kerogen-rich oil produced from oil shale rock. Shale oil, which is distinguished physically from heavy oil and tar sands, is an emerging petroleum source, and its potential was highlighted by the impressive production from the Bakken fields of North Dakota by the 2010s, which greatly boosted the state's petroleum output. (By 2015 North Dakota's daily petroleum production was approximately 1.2 million barrels, roughly 80 percent the amount produced per day by the country of Qatar, which is a member of Organization of the Petroleum Exporting Countries [OPEC].)

Tight oil is often light-gravity oil which is trapped in formations characterized by very low porosity and permeability. Tight oil production requires technologically complex drilling and completion methods, such as hydraulic fracturing (fracking) and other processes. (Completion is the practice of preparing the well and the equipment to extract petroleum.) The construction of horizontal wells with multi-fracturing completions is one of the most effective methods for recovering tight oil.

Formations containing light tight oil are dominated by siltstone containing quartz and other minerals such as dolomite and calcite. Mudstone may also be present. Since most formations look like shale oil on data logs (geologic reports), they are often referenced as shale. Higher-productivity tight oil appears to be linked to greater total organic carbon (TOC; the TOC fraction is the relative weight of organic carbon to kerogen in the sample) and greater shale thickness. Taken together, these factors may combine to create greater pore-pressure-related fracturing and more efficient extraction. For the most productive zones in the Bakken, TOC is estimated at greater than 40 percent, and thus it is considered to be a valuable source of hydrocarbons.

Other known commercial tight oil plays are located in Canada and Argentina. For example, Argentina's Vaca Muerta formation was expected to produce 350,000 barrels per well when fully exploited, but by the early 21st century only a few dozen wells had been drilled, which resulted in production of only a few hundred barrels per day. In addition, Russia's Bazhenov formation in west Siberia has 365 billion barrels of recoverable reserves, which is potentially greater than either Venezuela's or Saudi Arabia's proved conventional reserves.

Considering the commercial status of all unconventional petroleum resource plays, the most mature reside within the conterminous United States, where unconventional petroleum in the liquid, solid, and gaseous phases is efficiently extracted. For tight oil, further technological breakthroughs are expected to unlock the resource potential in a manner similar to how unconventional gas has been developed in the U.S.

Unconventional Natural Gas

Perhaps the most-promising advances for petroleum focus on unconventional natural gas. (Natural gas is a hydrocarbon typically found dissolved in oil or present as a cap for the oil in a petroleum deposit.) Six unconventional gas types—tight gas, deep gas, shale gas, coalbed methane, geopressurized zones, and Arctic and subsea hydrates—form the worldwide unconventional resource base. The scale of difference between conventional and unconventional reserves recoveries are commonly 30 percent to 1 percent, using tight gas as an example. In addition, the volume of the resource base is orders of magnitude higher; for example, 40 percent of all technically recoverable natural gas resources is attributable to shale gas. This total does not include tight gas, coalbed methane, or gas hydrates, nor does it include those shale gas resources that are believed to exist in unproven reserves in Russia and the Middle East.

World Distribution of Oil

Petroleum is not distributed evenly around the world. Slightly less than half of the world's proven reserves are located in the Middle East (including Iran but not North Africa). Following the Middle East are Canada and the United States, Latin America, Africa, and the region made up of Russia, Kazakhstan, and other countries that were once part of the Soviet Union.

The amount of oil and natural gas a given region produces is not always proportionate to the size of its proven reserves. For example, the Middle East contains approximately 50 percent of the world's proven reserves but accounts for only about 30 percent of global oil production (though this figure is still higher than in any other region). The United States, by contrast, lays claim to less than 2 percent of the world's proven reserves but produces roughly 16 percent of the world's oil.

Location of Reserves

Oil Fields

Two overriding principles apply to world petroleum production. First, most petroleum is contained in a few large fields, but most fields are small. Second, as exploration progresses, the average size of the fields discovered decreases, as does the amount of petroleum found per unit of exploratory drilling. In any region, the large fields are usually discovered first.

Since the construction of the first oil well in 1859, some 50,000 oil fields have been discovered. More than 90 percent of these fields are insignificant in their impact on world oil production. The two largest classes of fields are the supergiants, fields with 1 billion or more barrels of ultimately recoverable oil, and giants, fields with 500 million to 5 billion barrels of ultimately recoverable oil. Fewer than 40 supergiant oil fields have been found worldwide, yet these fields originally contained about one-half of all the oil so far discovered. The Arabian-Iranian sedimentary basin in the Persian Gulf region contains two-thirds of these supergiant fields. The remaining supergiants are distributed among the United States, Russia, Mexico, Libya, Algeria, Venezuela, China, and Brazil.

Although the semantics of what it means to qualify as a giant field and the estimates of recoverable reserves in giant fields differ between experts, the nearly 3,000 giant fields discovered—a figure which also includes the supergiants—account for 80 percent of the world's known recoverable oil. There are, in addition, approximately 1,000 known large oil fields that initially contained between 50 million and 500 million barrels. These fields account for some 14 to 16 percent of the world's known oil. Less than 5 percent of the known fields originally contained roughly 95 percent of the world's known oil.

Sedimentary Basins

Giant and supergiant petroleum fields and significant petroleum-producing basins of sedimentary rock are closely associated. In some basins, huge amounts of petroleum apparently have been generated because perhaps only about 10 percent of the generated petroleum is trapped and preserved. The Arabian-Iranian sedimentary basin is predominant because it contains more than 20 supergiant fields. No other basin has more than one such field. In 20 of the 26 most significant oil-containing basins, the 10 largest fields originally contained more than 50 percent of the known recoverable oil. Known world oil reserves are concentrated in a relatively small number of giant and supergiant fields in a few sedimentary basins.

Worldwide, approximately 600 sedimentary basins are known to exist. About 160 of these have yielded oil, but only 26 are significant producers, and 7 of these account for more than 65 percent of the total known oil. Exploration has occurred in another 240 basins, but discoveries of commercial significance have not been made.

Geologic Study and Exploration

Current geologic understanding can usually distinguish between geologically favourable and unfavourable conditions for oil accumulation early in the exploration cycle. Thus, only a relatively few exploratory wells may be necessary to indicate whether a region is likely to contain significant amounts of oil. Modern petroleum exploration is an efficient process. If giant fields exist, it is likely that most of the oil in a region will be found by the first 50 to 250 exploratory wells. This number may be exceeded if there is a much greater than normal amount of major prospects or if exploration drilling patterns are dictated by either political or unusual technological considerations. Thus, while undiscovered commercial oil fields may exist in some of the 240 explored but seemingly barren basins, it is unlikely that they will be of major importance since the largest are normally found early in the exploration process.

The remaining 200 basins have had little or no exploration, but they have had sufficient geologic study to indicate their dimensions, amount and type of sediments, and general structural character. Most of the underexplored (or frontier) basins are located in difficult environments, such as in polar regions, beneath salt layers, or within submerged continental margins. The larger sedimentary basins—those containing more than 833,000 cubic km (200,000 cubic miles) of sediments—account for some 70 percent of known world petroleum. Future exploration will have to involve the smaller basins as well as the more expensive and difficult frontier basins.

Status of the World Oil Supply

On several occasions—most notably during the oil crises of 1973–74 and 1978–79 and during the first half of 2008—the price of petroleum rose steeply. Because oil is such a crucial source of energy worldwide, such rapid rises in price spark recurrent debates about the accessibility of global supplies, the extent to which producers will be able to meet demand in the decades to come, and the potential for alternative sources of energy to mitigate concerns about energy supply and climate change issues related to the burning of fossil fuels.

How much oil does Earth have? The short answer to this question is that nobody knows. In its 1995 assessment of total world oil supplies, the U.S. Geological Survey (USGS) estimated that about 3 trillion barrels of recoverable oil originally existed on Earth and that about 710 billion barrels of that amount had been consumed by 1995. The survey acknowledged, however, that the total recoverable amount of oil could be higher or lower—3 trillion barrels was not a guess but an average of estimates based on different probabilities. This caveat notwithstanding, the USGS estimate was hotly disputed. Some experts said that technological improvements would create a situation in which much more oil would be ultimately recoverable, whereas others said that much less oil would be recoverable and that more than one-half of the world's original oil supply had already been consumed.

There is ambiguity in all such predictions. When industry experts speak of total "global oil reserves," they refer specifically to the amount of oil that is thought to be recoverable, not the total amount remaining on Earth. What is counted as "recoverable," however, varies from estimate to estimate. Analysts make distinctions between "proven reserves"—those that can be demonstrated as recoverable with reasonable certainty, given existing economic and technological conditions—and reserves that may be recoverable but are more speculative. The Oil & Gas Journal, estimated in late 2007 that the world's proven reserves amounted to roughly 1.3 trillion barrels. To put this number in context, the world's population consumed about 30 billion barrels of oil in 2007. At this rate of consumption, disregarding any new reserves that might be found, the world's proven reserves would be depleted in about 43 years. However, because of advancements in exploration and unconventional oil extraction, estimates of the world's proven oil reserves had risen to approximately 1.7 trillion barrels by 2015.

By any estimation, it is clear that Earth has a finite amount of oil and that global demand is expected to increase. In 2007 the National Petroleum Council, an advisory committee to the U.S. Secretary of Energy, projected that world demand for oil would rise from 86 million barrels per day to as much as 138 million barrels per day in 2030. Yet experts remain divided on whether the world will be able to supply so much oil. Some argue that the world has reached "peak oil"—its peak rate of oil production. The controversial theory behind this argument draws on studies that show how production from individual oil fields and from oil-producing regions has tended to increase to a point in time and then decrease thereafter. "Peak-oil theory" suggests that once global peak oil has been reached, the rate of oil production in the world will progressively decline, with severe economic consequences to oil-importing countries.

A more widely accepted view is that through the early 21st century at least, production capacity will be limited not by the amount of oil in the ground but by other factors, such as geopolitics or economics. One concern is that growing dominance by nationalized oil companies, as opposed to independent oil firms, can lead to a situation in which countries with access to oil reserves will limit production for political or economic gain. A separate concern is that nonconventional sources of oil—such as oil sand reserves, oil shale deposits, or reserves that are found under very deep water—will be significantly more expensive to produce than conventional crude oil unless new technologies are developed that reduce production costs.

Natural Gas

Natural gas is a colourless highly flammable gaseous hydrocarbon consisting primarily of methane and ethane. It is a type of petroleum that commonly occurs in association with crude oil. A fossil fuel, natural gas is used for electricity generation, heating, and cooking and as a fuel for certain vehicles. It is important as a chemical feedstock in the manufacture of plastics and is necessary for a wide array of other chemical products, including fertilizers and dyes.

Natural gas is often found dissolved in oil at the high pressures existing in a reservoir, and it can be present as a gas cap above the oil. In many instances it is the pressure of natural gas exerted upon the subterranean oil reservoir that provides the drive to force oil up to the surface. Such natural gas is known as associated gas; it is often considered to be the gaseous phase of the crude oil and usually contains some light liquids such as propane and butane. For this reason, associated gas is sometimes called "wet gas." There are also reservoirs that contain gas and no oil. This gas is termed nonassociated gas. Nonassociated gas, coming from reservoirs that are not connected with any known source of liquid petroleum, is "dry gas."

The first discoveries of natural gas seeps were made in Iran between 6000 and 2000 BCE. Many early writers described the natural petroleum seeps in the Middle East, especially in the Baku region of what is now Azerbaijan. The gas seeps, probably first ignited by lightning, provided the fuel for the "eternal fires" of the fire-worshipping religion of the ancient Persians.

The use of natural gas was mentioned in China about 900 BCE. It was in China in 211 BCE that the first known well was drilled for natural gas, to reported depths of 150 metres (500 feet). The Chinese drilled their wells with bamboo poles and primitive percussion bits for the express purpose of searching for gas in limestones dating to the Late Triassic Epoch (about 237 million to 201.3 million years ago) in an anticline (an arch of stratified rock) west of modern Chongqing. The gas was burned to dry the rock salt found interbedded in the limestone. Eventually wells were drilled to depths approaching 1,000 metres (3,300 feet), and more than 1,100 wells had been drilled into the anticline by 1900.

Natural gas was unknown in Europe until its discovery in England in 1659, and even then it did not come into wide use. Instead, gas obtained from carbonized coal (known as town gas) became the primary fuel for illuminating streets and houses throughout much of Europe from 1790 on.

In North America the first commercial application of a petroleum product was the utilization of natural gas from a shallow well in Fredonia, New York, in 1821. The gas was distributed through a small-bore lead pipe to consumers for lighting and cooking.

Improvements in Gas Pipelines

Throughout the 19th century the use of natural gas remained localized because there was no way to transport large quantities of gas over long distances. Natural gas remained on the sidelines of industrial development, which was based primarily on coal and oil. An important breakthrough in gas-transportation technology occurred in 1890 with the invention of leakproof pipeline coupling. Nonetheless, materials and construction techniques remained so cumbersome that gas could not be used more than 160 km (100 miles) from a source of supply. Thus, associated gas was mostly flared (i.e., burned at the wellhead), and nonassociated gas was left in the ground, while town gas was manufactured for use in the cities.

Long-distance gas transmission became practical during the late 1920s because of further advances in pipeline technology. From 1927 to 1931 more than 10 major transmission systems were constructed in the United States. Each of these systems was equipped with pipes having diameters of approximately 50 cm (20 inches) and extended more than 320 km (200 miles). Following World War II, a large number of even longer pipelines of increasing diameter were constructed. The fabrication of pipes having a diameter of up to 150 cm (60 inches) became possible. Since the early 1970s the longest gas pipelines have had their origin in Russia. For example, in the 1960s and '70s the 5,470-km- (3,400-mile-) long Northern Lights pipeline was built across the Ural Mountains and some 700 rivers and streams, linking eastern Europe with the West Siberian gas fields on the Arctic Circle. As a result, gas from the Urengoy field, the world's largest, is now transported to eastern Europe and then on to western Europe for consumption. Another gas pipeline, shorter but also of great engineering difficulty, was the 50-cm (20-inch) Trans-Mediterranean Pipeline, which during the 1970s and '80s was constructed between Algeria and Sicily. The sea is more than 600 metres (2,000 feet) deep along some parts of that route.

Natural gas pipeline in Temane.

Natural Gas as a Premium Fuel

As recently as 1960, associated gas was a nuisance by-product of oil production in many areas of the world. The gas was separated from the crude oil stream and eliminated as cheaply as possible, often by flaring (burning it off). Only after the crude oil shortages of the late 1960s and early '70s did natural gas become an important world energy source.

Even in the United States the home-heating market for natural gas was limited until the 1930s, when town gas began to be replaced by abundant and cheaper supplies of natural gas, which contained twice the heating value of its synthetic predecessor. Also, when natural gas burns completely, carbon dioxide and water are normally formed. The combustion of gas is relatively free of soot, carbon monoxide, and the nitrogen oxides associated with the burning of other fossil fuels. In addition, sulfur dioxide emissions, another major air pollutant, are almost nonexistent. As a consequence, natural gas is often a preferred fuel for environmental reasons, and it is supplanting coal as a fuel for electric power plants in many parts of the world.

Composition and Properties of Natural Gas

Hydrocarbon Content

Natural gas is a hydrocarbon mixture consisting primarily of saturated light paraffins such as methane and ethane, both of which are gaseous under atmospheric conditions. The mixture also may contain other hydrocarbons, such as propane, butane, pentane, and hexane. In natural gas reservoirs even the heavier hydrocarbons occur for the most part in gaseous form because of the higher pressures. They usually liquefy at the surface (at atmospheric pressure) and are produced separately as natural gas liquids (NGLs), either in field separators or in gas processing plants. Once separated from the gas stream, the NGLs can be further separated into fractions, ranging from the heaviest condensates (hexanes, pentanes, and butanes) through liquefied petroleum gas (LPG;

essentially butane and propane) to ethane. This source of light hydrocarbons is especially prominent in the United States, where natural gas processing provides a major portion of the ethane feedstock for olefin manufacture and the LPG for heating and commercial purposes.

Nonhydrocarbon Content

Other gases that commonly occur in association with the hydrocarbon gases are nitrogen, carbon dioxide, hydrogen, and such noble gases as helium and argon. Nitrogen and carbon dioxide are noncombustible and may be found in substantial proportions. Nitrogen is inert, but, if present in significant amounts, it reduces the heating value of the mixture; it must therefore be removed before the gas is suitable for the commercial market. Carbon dioxide is removed in order to raise heating value, reduce volume, and sustain even combustion properties.

Often natural gases contain substantial quantities of hydrogen sulfide or other organic sulfur compounds. In this case, the gas is known as "sour gas." Sulfur compounds are removed in processing, as they are toxic when breathed, are corrosive to plant and pipeline facilities, and are serious pollutants if burned in products made from sour gas. However, after sulfur removal a minute quantity of a noxious mercaptan odorant is always added to commercial natural gas in order to ensure the rapid detection of any leakage that may occur in transport or use.

Because natural gas and formation water occur together in the reservoir, gas recovered from a well contains water vapour, which is partially condensed during transmission to the processing plant.

Thermal and Physical Properties

Commercial natural gas stripped of NGL and sold for heating purposes usually contains 85 to 90 percent methane, with the remainder mainly nitrogen and ethane. It usually has a calorific, or heating, value of approximately 38 megajoules (MJ; million joules) per cubic metre or about 1,050 British thermal units (BTUs) per cubic foot of gas.

Methane is colourless, odourless, and highly flammable. However, some of the associated gases in natural gas, especially hydrogen sulfide, have a distinct and penetrating odour, and a few parts per million are sufficient to impart a decided odour to natural gas.

Processing and Transport of Natural Gas

Measurement Systems

The amounts of gas accumulated in a reservoir, as well as produced from wells and transported through pipelines, are measured by volume, calculated in either cubic metres or cubic feet. The calculations are made with reference to the volume occupied by the gas at standard atmospheric pressure (i.e., 760 mm of mercury, or 14.7 pounds per square inch) and at a temperature of 15 °C (60 °F). Since gas in the reservoir is compressed by the high pressures exerted underground, it expands upon reaching the surface and thus occupies more space. However, since its volume is calculated in reference to standard conditions of temperature and pressure, this expansion does not constitute an increase in the amount of gas produced. Natural gas reserves are usually measured in billions and trillions of cubic metres (bcm and tcm) or in billions and trillions of cubic feet (bcf and tcf). Volumes produced on a daily basis at wells are frequently measured in thousands and millions of cubic metres (Mcm and MMcm) or in thousands and millions of cubic feet (Mcf and MMcf). By tradition the natural gas industry uses the Roman numeral M to designate 1,000 and MM (1,000 × 1,000) to denote one million.

On the market, natural gas is usually bought and sold not by volume but by calorific value, noted above as approximately 38 MJ per cubic metre or about 1,050 BTUs per cubic foot. These units are frequently abbreviated as MJ/m³ and BTU/ft³. In practice, purchases of natural gas are usually denoted in much larger units, such as GJ (gigajoules, billions of joules) and MMBTUs (millions of BTUs).

In the British Imperial system, 1 MMBTU is conveniently equivalent to roughly 1,000 cubic feet of natural gas. Another unit frequently used is the therm, which is equivalent to 100,000 BTUs or roughly 100 cubic feet of gas. The price of natural gas is frequently cited per therm, per MMBTU, or per GJ.

Field Processing

Sometimes field-production gas is high enough in methane content that it can be piped directly to customers without processing. Most often, however, the gas contains unacceptable levels of higher-weight hydrocarbon liquids as well as impurities, and it is available only at very low pressures. For these reasons, field gas is usually processed through multiple stages of compression to remove liquids and impurities and to reduce the temperature of the fluid in order to conserve the power requirements of compressor stations along the transport pipeline.

Dehydration

In a simple compression gas-processing plant, field gas is charged to an inlet scrubber, where entrained liquids are removed. The gas is then successively compressed and cooled. As the pressure is increased and the temperature reduced, water vapour in the gas condenses. If liquid forms in the coolers, the gas may be at its dew point with respect to water or hydrocarbons. This may result in the formation of icelike gas hydrates, which can cause difficulty in plant operation and must be prevented from forming in order to avoid problems in subsequent transportation. Hydrate prevention is accomplished by injecting a glycol solution into the process stream to absorb any dissolved water. The dehydrated gas continues through the processing stream, and the glycol solution, containing absorbed water, is heated to evaporate the water and is then reused.

Another dehydration method involves passing the wet gas through a succession of towers packed with a solid desiccant material. Water dissolved in the gas is adsorbed onto the desiccant, and the dry gas emerges for further processing.

Recovery of Hydrocarbon Liquids

If market economics warrant the recovery of NGLs from the gas stream, a more complex absorption and fractionation plant may be required. The compressed raw gas is processed in admixture with a liquid hydrocarbon, called lean oil, in an absorber column, where heavier components in the gas are absorbed in the lean oil. The bulk of the gas is discharged from the top of the absorber as residue gas (usually containing 95 percent methane) for subsequent treatment to remove sulfur and other impurities. The heavier components leave with the bottoms liquid stream, now called rich oil, for further processing in a distillation tower to remove ethane for plant fuel or petrochemical feedstock and to recover the lean oil. Some gas-processing plants may contain additional distilling columns for further separation of the NGL into propane, butane, and heavier liquids.

Many older gas-absorption plants were designed to operate at ambient temperature, but some more modern facilities employ refrigeration to lower processing temperatures and increase the absorption efficiency. An even more efficient process, especially for extracting ethane, is known as cryogenic expansion. In this process cooled gas is blown by a powerful turbine into an expansion chamber, where the vapour pressure of the gas is reduced and its temperature further lowered to −84 °C (−120 °F). At this temperature methane is still a gas, but the heavier hydrocarbons condense and are recovered.

Sweetening

Sour gas is sweetened, or purified of its sulfur compounds, by treatment with ethanolamine, a liquid absorbent that acts much like the glycol solution in dehydration. After bubbling through the liquid, the gas emerges almost entirely stripped of sulfur. The ethanolamine is processed for removal of the absorbed sulfur and is reused.

Transport

The growth of the natural gas industry has largely depended on the development of efficient pipeline systems. The first metal pipeline was constructed between Titusville and Newton, Pennsylvania, in 1872. This 2.5-inch- (6.4-cm-) diameter cast-iron system supplied some 250 residential customers with natural gas at a pressure of about 80 pounds per square inch (psi), or 550 kilopascals (KPa). By the early 21st century more than 500,000 km (300,000 miles) of main transmission pipelines and 3.4 million km (2.1 million miles) of smaller distribution pipelines were operating in the United States, delivering more than 672 bcm (24 tcf) of natural gas per year to some 70 million customers. Russia, the world's largest gas exporter, was operating more than 160,000 km (100,000 miles) of transmission pipelines with the capacity to transport more than 600 bcm (21 tcf) of natural gas per year.

Compressor: A compressor station on a natural gas pipeline.

Modern gas pipelines are built in numerous sizes, depending on their use, with diameters ranging from 15 cm (6 inches) for feeder lines to diameters such as 60, 106, and 122 cm (24, 42, and 48 inches) for transmission pipelines. The biggest Russian main lines have diameters as high as 140 cm (56 inches). Large transmission pipelines operate at pressures up to about 8 megapascals

(MPa), or more than 1,000 psi. (In parts of the world that use the metric system, pipeline pressures are also measured in bars. One bar equals 100 KPa, so 8 MPa, or 8,000 KPa, is 80 bars.) Automated compressor stations are located approximately every 100 km (60 miles) along the pipelines to boost system pressure and overcome friction losses in transit.

The presence of natural gas fields in areas of the world far from market destinations has given rise to an efficient means of long-distance oceanic transport. Since liquefied natural gas (LNG) occupies only 0.16 percent (1/600) of the gaseous volume, an international trade has naturally developed in LNG. Modern liquefaction plants employ autorefrigerated cascade cycles, in which the gas is stripped of carbon dioxide, dried, and then subjected to a series of compression-expansion steps during which it is cooled to liquefaction temperature (approximately −160 °C [−260 °F]). The compression power requirement is usually supplied by consuming a portion of the available gas. After liquefaction the gas is transported in specially designed and insulated tankers to the consuming port, where it is stored in refrigerated tanks until required. Regasification requires a source of heat to convert the liquid back into vapour. Often a low-cost method is followed, such as exchanging heat with a large volume of nearby seawater. All methods of liquefaction, transport, and regasification involve a significant energy loss, which can approach 25 percent of the original energy content of the gas.

Applications

The largest single application for natural gas is as a fuel for electric power generation. Power generation is followed by industrial, domestic, and commercial uses—mainly as a source of energy but also, for instance, as a feedstock for chemical products. Several specialized applications have developed over the years. The clean-burning characteristics of natural gas have made it a frequent choice as a nonpolluting transportation fuel, though it does emit the greenhouse gas carbon dioxide. Many buses and commercial automotive fleets now operate on compressed natural gas. Carbon black, a pigment of colloidal dimensions, is made by burning natural gas with a limited supply of air and depositing the soot on a cool surface. It is an important ingredient in dyes and inks and is used in rubber compounding operations.

More than half of the world's ammonia supply is manufactured via a catalytic process that uses hydrogen derived from methane. Ammonia is used directly as a plant food or converted into a variety of chemicals such as hydrogen cyanide, nitric acid, urea, and a range of fertilizers.

A wide array of other chemical products can be made from natural gas by a controlled oxidation process—for example, methanol, propanol, and formaldehyde, which serve as basic materials for a wide range of other chemical products. Methanol can be used as a gasoline additive or gasoline substitute. In addition, methyl tertiary butyl ether (MTBE), an oxygenated fuel additive added to gasoline in order to raise its octane number, is produced via chemical reaction of methanol and isobutylene over an acidic ion-exchange resin.

Organic Formation Process

Natural gas is more ubiquitous than oil. It is derived from both land plants and aquatic organic matter and is generated above, throughout, and below the oil window. Thus, all source rocks have the potential for gas generation. Many of the source rocks for significant gas deposits appear to be associated with the worldwide occurrence of coal dated to Carboniferous and Early Permian times (roughly 358.9 million to 273 million years ago).

Biological Stage

During the immature, or biological, stage of petroleum formation, biogenic methane (often called marsh gas) is produced as a result of the decomposition of organic material by the action of anaerobic microbes. These microorganisms cannot tolerate even traces of oxygen and are also inhibited by high concentrations of dissolved sulfate. Consequently, biogenic gas generation is confined to certain environments that include poorly drained swamps and bays, some lake bottoms, and marine environments beneath the zone of active sulfate reduction. Gas of predominantly biogenic origin is thought to constitute more than 20 percent of the world's gas reserves.

The mature stage of petroleum generation, which occurs at depths of about 750 to 5,000 metres (2,500 to 16,000 feet), includes the full range of hydrocarbons that are produced within the oil window. Often significant amounts of thermal methane gas are generated along with the oil. Below 2,900 metres (9,500 feet), primarily wet gas (gas containing liquid hydrocarbons) is formed.

Thermal Stage

In the postmature stage, below about 5,000 metres (16,000 feet), oil is no longer stable, and the main hydrocarbon product is thermal methane gas. The thermal gas is the product of the cracking of the existing liquid hydrocarbons. Those hydrocarbons with a larger chemical structure than that of methane are destroyed much more rapidly than they are formed. Thus, in the sedimentary basins of the world, comparatively little oil is found below 5,000 metres. The deep basins with thick sequences of sedimentary rocks, however, have the potential for deep gas production.

Inorganic Formation

Some methane may have been produced by inorganic processes. The original source of Earth's carbon was the cosmic debris from which the planet formed. If meteorites are representative of this debris, the carbon could have been supplied in comparatively high concentrations as hydrocarbons, such as are found in the carbonaceous chondrite type of meteorites. Continuous outgassing of these hydrocarbons may be taking place from within Earth, and some may have accumulated as abiogenic gas deposits without having passed through an organic phase. In the event of widespread outgassing, however, it is likely that abiogenic gas would be too diffuse to be of commercial interest. Significant accumulations of inorganic methane have yet to be found.

The helium and some of the argon found in natural gas are products of natural radioactive disintegration. Helium derives from radioisotopes of thorium and the uranium family, and argon derives from potassium. It is probably coincidental that helium and argon sometimes occur with natural gas; in all likelihood, the unrelated gases simply became caught in the same trap.

Geologic Environment

Like oil, natural gas migrates and accumulates in traps. Oil accumulations contain more recoverable energy than gas accumulations of similar size, even though the recovery of gas is a more efficient process than the recovery of oil. This is due to the differences in the physical and chemical properties of gas and oil. Gas displays initial low concentration and high dispersibility, making adequate cap rocks very important.

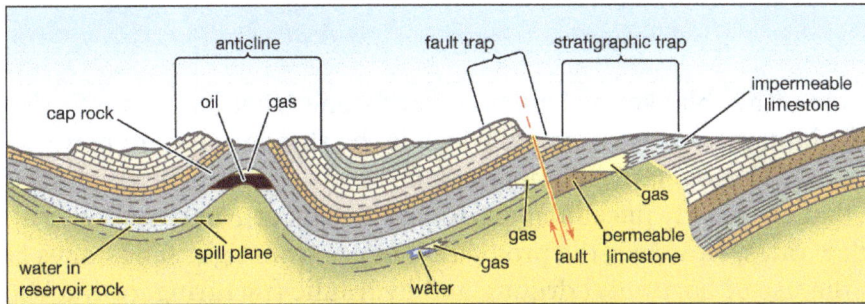

Principal types of petroleum traps.

Natural gas can be the primary target of either deep or shallow drilling because large gas accumulations form above the oil window as a result of biogenic processes and thermal gas occurs throughout and below the oil window. In most sedimentary basins the vertical potential (and sediment volume) available for gas generation exceeds that of oil. About a quarter of the known major gas fields are related to a shallow biogenic origin, but most major gas fields are located at intermediate or deeper levels where higher temperatures and older reservoirs (often carbonates sealed by evaporites) exist.

Conventional Gas Reservoirs

Gas reservoirs differ greatly, with different physical variations affecting reservoir performance and recovery. In a natural gas (single-phase) reservoir it should be possible to recover nearly all of the in-place gas by dropping the pressure sufficiently. If the pressure is effectively maintained by the encroachment of water in the sedimentary rock formation, however, some of the gas will be lost to production by being trapped by capillarity behind the advancing water front. Therefore, in practice, only about 80 percent of the in-place gas can be recovered. On the other hand, if the pressure declines, there is an economic limit at which the cost of compression exceeds the value of the recovered gas. Depending on formation permeability, actual gas recovery can be as high as 75 to 80 percent of the original in-place gas in the reservoir. Associated gas is produced along with the oil and is separated at the surface.

Unconventional Gas Reservoirs

Substantial amounts of gas have accumulated in geologic environments that differ from conventional petroleum traps. This gas is termed unconventional gas and occurs in "tight" (i.e., relatively impermeable) sandstones, in joints and fractures or absorbed into the matrix of shales, and in coal seams. In addition, large amounts of gas are locked into methane hydrates in cold polar and undersea regions, and gas is also present dissolved or entrained in hot geopressured formation waters.

Unconventional gas sources are unconventional only in the sense that, given current economic conditions and states of technology, they are more expensive to exploit and may produce at much slower rates than conventional gas fields. However, as technology changes or as conventional sources become relatively expensive, some unconventional gas becomes easier and relatively cheaper to produce in quantities that can fully complement conventional gas production. Such has been the case with tight gas, shale gas, and coal-bed methane.

Tight Gas

Tight gas occurs in either blanket or lenticular sandstones that have an effective permeability of less than one millidarcy (or 0.001 darcy, which is the standard unit of permeability of a substance to fluid flow). These relatively impermeable sandstones are reservoirs for considerable amounts of gas that are mostly uneconomical to produce by conventional vertical wells because of low natural flow rates. However, the production of gas from tight sandstones has been greatly enhanced by the use of horizontal drilling and hydraulic fracturing, or fracking, techniques, which create large collection areas in low-permeability formations through which gas can flow to a producing well.

Shale Gas

Shale gas was generated from organic mud deposited at the bottom of ancient bodies of water. Subsequent sedimentation and the resultant heat and pressure transformed the mud into shale and also produced natural gas from the organic matter contained in it. Over long spans of geologic time, some of the gas migrated to adjacent sandstones and was trapped in them, forming conventional gas accumulations. The rest of the gas remained locked in the nonporous shale. In the past the production of shale gas was generally too slow to be profitable, but now wells can be drilled horizontally for long distances through the shale beds, and the formations can be stimulated by hydraulic fracturing to enhance gas production greatly. About 25 percent of the gas produced in the United States comes from shales, and that proportion is expected to rise to 50 percent before the mid-21st century.

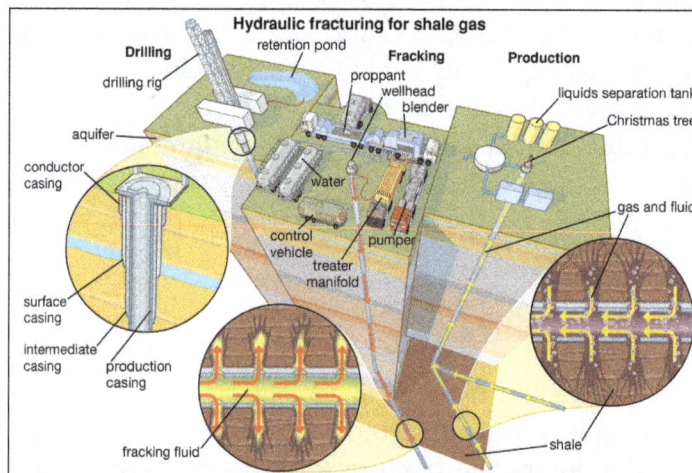

Three steps in the extraction of shale gas: drilling a borehole into the shale formation and lining it with pipe casing; fracking, or fracturing, the shale by injecting fluid under pressure; and producing gas that flows up the borehole, frequently accompanied by liquids.

Coal-bed Methane

Considerable quantities of methane are trapped within coal seams. Although much of the gas that formed during the initial coalification process is lost to the atmosphere, a significant portion remains as free gas in the joints and fractures of the coal seam; in addition, large quantities of gas are adsorbed on the internal surfaces of the micropores within the coal itself. This gas can be accessed by drilling wells into the coal seam and pumping out large quantities of water that saturate the seam. Removing the water lowers the pressure in the seam, allowing the adsorbed methane to desorb and migrate as free gas into fractures in the coal; from there it enters the wellbore and is brought to the surface. Since coal is relatively impermeable, the existing fracture systems of seams that contain rich reserves of methane are sometimes stimulated by fracking in a manner similar to shales and tight sandstones. Coal-bed gas accounts for almost 10 percent of total gas output in the United States, and it is becoming an important source of natural gas in other regions of the world as well.

Geopressured Fluids and Methane Hydrates

Geopressured reservoirs exist throughout the world in deep, geologically young sedimentary basins in which the formation fluids (which usually occur in the form of a brine) bear a part of the overburden load. The fluid pressures can become quite high, sometimes almost double the normal hydrostatic gradient. In many cases the geopressured fluids also become hotter than normally pressured fluids, because the heat flow to the surface is impeded by insulating layers of impermeable shales and clays. Geopressured fluids have been found to be saturated with 0.84 to 2.24 cubic metres of natural gas per 0.159 cubic metre of brine, or 30 to 80 cubic feet of gas per barrel. To produce this gas, high flow rates of the hot geopressured fluids must be maintained from formations of high porosity and permeability. Because very large amounts of formation water must be produced to recover commercial quantities of the associated gas, there is no commercial gas production known to be derived from a geopressured deposit.

Enormous quantities of natural gas are estimated to be locked up in so-called methane hydrates, which are unusual molecular structures in which single methane molecules are encased in icy cagelike lattices of water molecules. Methane hydrates are found beneath the permafrost in polar regions and also in the ocean bed along the outer edges of continental shelves. In both of these environments, very specific combinations of pressure and temperature produce conditions that allow methane to migrate into reservoirs containing water and for the two species to form the hydrate structures. Methane hydrates have been found in sandstones from polar regions and in sand and mud sediments from continental margins. Techniques for extracting the methane in an economically viable and environmentally sustainable manner are under exploration. One possibility is to drill into a hydrate-rich formation and reduce the pressure in the surrounding rock sufficiently to release the methane from the water lattice. Another is to pump carbon dioxide into the formation. The carbon dioxide molecules would replace the methane molecules in the lattice structure, releasing the methane for extraction through a borehole. Any extraction technology would have to be carefully designed around the extremely sensitive polar ecosystems and marine ecosystems where the reserves are located.

Nuclear Power

Nuclear power is the use of nuclear reactions that release nuclear energy to generate heat, which most frequently is then used in steam turbines to produce electricity in a nuclear power plant. As a nuclear technology, nuclear power can be obtained from nuclear fission, nuclear decay and nuclear fusion reactions. Presently, the vast majority of electricity from nuclear power is produced by nuclear fission of uranium and plutonium. Nuclear decay processes are used in niche applications such as radioisotope thermoelectric generators. Generating electricity from fusion power remains at the focus of international research.

Civilian nuclear power supplied 2,488 terawatt hours (TWh) of electricity in 2017, equivalent to about 10% of global electricity generation, and was the second largest low-carbon power source after hydroelectricity. As of April 2018, there are 449 civilian fission reactors in the world, with a combined electrical capacity of 394 gigawatt (GW). As of 2018, there are 58 power reactors under construction and 154 reactors planned, with a combined capacity of 63 GW and 157 GW, respectively. As of January 2019, 337 more reactors were proposed. Most reactors under construction are generation III reactors in Asia.

Since its commercialization in the 1970s, nuclear power has prevented about 1.84 million air pollution-related deaths and the emission of about 64 billion tonnes of carbon dioxide equivalent that would have otherwise resulted from the burning of fossil fuels.

There is a debate about nuclear power. Proponents, such as the World Nuclear Association and Environmentalists for Nuclear Energy, contend that nuclear power is a safe, sustainable energy source that reduces carbon emissions. Opponents, such as Greenpeace and NIRS, contend that nuclear power poses many threats to people and the environment.

Accidents in nuclear power plants include the Chernobyl disaster in the Soviet Union in 1986, the Fukushima Daiichi nuclear disaster in Japan in 2011, and the more contained Three Mile Island accident in the United States in 1979. There have also been some nuclear submarine accidents. Nuclear reactors have caused the lowest number of fatalities per unit of energy generated when compared to fossil fuels and hydropower. Coal, petroleum, natural gas and hydroelectricity each have caused a greater number of fatalities per unit of energy, due to air pollution and accidents.

Collaboration on research and development towards greater efficiency, safety and recycling of spent fuel in future generation IV reactors presently includes Euratom and the co-operation of more than 10 permanent member countries globally.

The Hanul Nuclear Power Plant in South Korea, presently the second largest in the world by output, with six operating power reactors. Two additional indigenously-designed APR-1400 generation-III reactors are under construction. South Korea exported the APR design to the United Arab Emirates, where four of these reactors are under construction at Barakah nuclear power plant.

As of 2018, there are over 150 nuclear reactors planned including 50 under construction. However, while investment on upgrades of existing plant and life-time extensions continues, investment in new nuclear is declining, reaching a 5-year-low in 2017.

In 2016, the U.S. Energy Information Administration projected for its "base case" that world nuclear power generation would increase from 2,344 terawatt hours (TWh) in 2012 to 4,500 TWh in 2040. Most of the predicted increase was expected to be in Asia.

The future of nuclear power varies greatly between countries, depending on government policies. Some countries, most notably, Germany, have adopted policies of nuclear power phase-out. At the same time, some Asian countries, such as China and India, have committed to rapid expansion of nuclear power. Many other countries, such as the United Kingdom and the United States, have policies in between. Japan generated about 30% of its electricity from nuclear power before the Fukushima accident. In 2015 the Japanese government committed to the aim of restarting its fleet of 40 reactors by 2030 after safety upgrades, and to finish the construction of the Generation III Ōma Nuclear Power Plant. This would mean that approximately 20% of electricity would come from nuclear power by 2030. As of 2018, some reactors have restarted commercial operation following inspections and upgrades with new regulations. While South Korea has a large nuclear power industry, the new government in 2017, influenced by a vocal anti-nuclear movement, committed to halting nuclear development after the completion of the facilities presently under construction.

The Generation IV roadmap. Nuclear Energy Systems Deployable no later than 2030 and offering significant advances in sustainability, safety and reliability, and economics.

The nuclear power industry in some western nations have a history of construction delays, cost overruns, plant cancellations, and nuclear safety issues, despite significant government subsidies and support. These problems are related to very strict safety requirements, uncertain regulatory environment, slow rate of construction, and large stretches of time with no nuclear construction and consequent loss of know-how. Commentators therefore argue that new nuclear is impractical in western countries because of popular opposition, regulatory uncertainty, soft demand for multiple reactor units and high costs. The bankruptcy of Westinghouse in March 2017 due to US$9 billion of losses from the halting of construction at Virgil C. Summer Nuclear Generating Station, in the U.S. is considered an advantage for eastern companies, for the future export and design of nuclear fuel and reactors. In 2016, Greenpeace and the wind industry company Ecotricity criticized the high cost of the Hinkley Point C nuclear power station and threatened to take action in British or French courts or lodge a complaint with the European Commission, in order to trigger an investigation, which they said could last as long as a year.

The greatest new build activity is occurring in Asian countries like South Korea, India and China. In January 2019, China had 45 reactors in operation, 13 under construction, and plans to build 43 more, which would make it the world's largest generator of nuclear electricity.

Blue light from Cherenkov radiation being produced near the core of the Fission powered Advanced Test Reactor. A facility taking part in the Advanced Fuel Cycle Initiative, to transmute certain actinides into fuel, that would be able to be used in commercial light water reactors, reducing a number of the security hazards of, what is all presently considered "waste".

In 2016 the BN-800 sodium cooled fast reactor in Russia, began commercial electricity generation, while plans for a BN-1200 were initially conceived the future of the fast reactor program in Russia awaits the results from MBIR, an under construction multi-loop Generation IV research facility for testing the chemically more inert lead, lead-bismuth and gas coolants, it will similarly run on recycled MOX (mixed uranium and plutonium oxide) fuel. An on-site pyrochemical processing, closed fuel-cycle facility, is planned, to recycle the spent fuel/"waste" and reduce the necessity for a growth in uranium mining and exploration. In 2017 the manufacture program for the reactor commenced with the facility open to collaboration under the "International Project on Innovative Nuclear Reactors and Fuel Cycle", it has a construction schedule, that includes an operational start in 2020. As planned, it will be the world's most-powerful research reactor.

Extending Plant Lifetimes

As of 2019 the cost of extending plant lifetimes is competitive with other electricity generation technologies, including new solar and wind projects. In the United States, licenses of almost half of the operating nuclear reactors have been extended to 60 years. The U.S. NRC and the U.S. Department of Energy have initiated research into Light water reactor sustainability which is hoped will lead to allowing extensions of reactor licenses beyond 60 years, provided that safety can be maintained, to increase energy security and preserve low-carbon generation sources. Research into nuclear reactors that can last 100 years, known as Centurion Reactors, is being conducted.

Nuclear Power Station

Just as many conventional thermal power stations generate electricity by harnessing the thermal energy released from burning fossil fuels, nuclear power plants convert the energy released from the nucleus of an atom via nuclear fission that takes place in a nuclear reactor. When a neutron hits the nucleus of a uranium-235 or plutonium atom, it can split the nucleus into two smaller nuclei. The reaction is called nuclear fission. The fission reaction releases energy and neutrons. The released neutrons can hit other uranium or plutonium nuclei, causing new fission reactions, which release more energy and more neutrons. This is called a chain reaction. The reaction rate is controlled by control rods that absorb excess neutrons. The controllability of nuclear reactors depends on the fact that a small fraction of neutrons resulting from fission are delayed. The time delay between the fission and the release of the neutrons slows down changes in reaction rates and gives time for moving the control rods to adjust the reaction rate.

Pressurized water reactor in operation.

A fission nuclear power plant is generally composed of a nuclear reactor, in which the nuclear reactions generating heat take place; a cooling system, which removes the heat from inside the reactor; a steam turbine, which transforms the heat in mechanical energy; an electric generator, which transform the mechanical energy into electrical energy.

Life Cycle of Nuclear Fuel

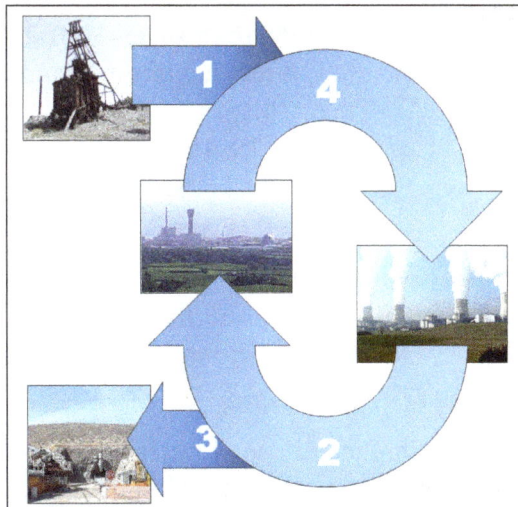

The nuclear fuel cycle begins when uranium is mined, enriched, and manufactured into nuclear fuel, (1) which is delivered to a nuclear power plant. After usage in the power plant, the spent fuel is delivered to a reprocessing plant (2) or to a final repository (3) for geological disposition. In repro-cessing 95% of spent fuel can potentially be recycled to be returned to (4) usage in a power plant.

A nuclear reactor is only part of the fuel life-cycle for nuclear power. The process starts with mining. Uranium mines are underground, open-pit, or in-situ leach mines. The uranium ore is extracted, usually converted into a stable and compact form such as yellowcake, and then transported to a processing facility. Here, the yellowcake is converted to uranium hexafluoride, which is then generally enriched using various techniques. Some reactor designs can also use natural uranium without enrichment. The enriched uranium, containing more than the natural 0.7% uranium-235, is generally used to make rods of the proper composition and geometry for the particular reactor that the fuel is destined for. In modern light-water reactors the fuel rods will spend about 3 operational cycles (typically 6 years total now) inside the reactor, generally until about 3% of their uranium has been fissioned, then they will be moved to a spent fuel pool where the short lived isotopes generated by fission can decay away. After about 5 years in a spent fuel pool the spent fuel is radioactively and thermally cool enough to handle, and can be moved to dry storage casks or reprocessed.

Conventional Fuel Resources

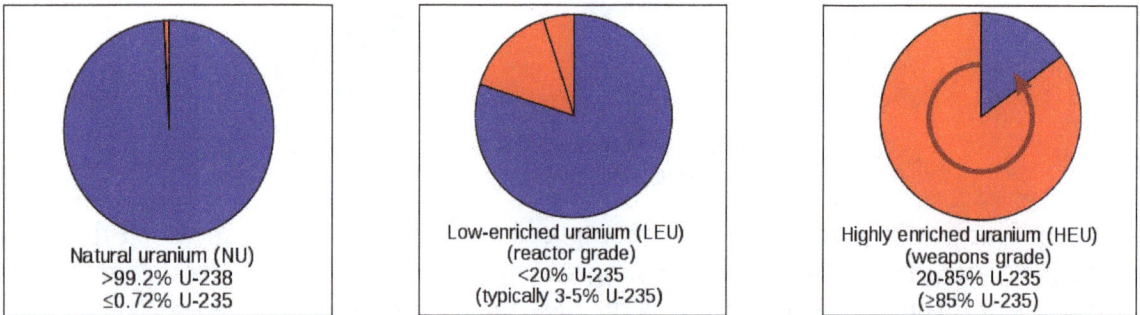

Natural uranium (NU) >99.2% U-238 ≤0.72% U-235	Low-enriched uranium (LEU) (reactor grade) <20% U-235 (typically 3-5% U-235)	Highly enriched uranium (HEU) (weapons grade) 20-85% U-235 (≥85% U-235)

Proportions of the isotopes uranium-238 (blue) and uranium-235 (red) found in natural uranium and in enriched uranium for different applications. Light water reactors use 3-5% enriched uranium, while CANDU reactors work with natural uranium.

Uranium is a fairly common element in the Earth's crust: it is approximately as common as tin or germanium, and is about 40 times more common than silver. Uranium is present in trace concentrations in most rocks, dirt, and ocean water, but is generally economically extracted only where it is present in high concentrations. As of 2011 the world's known resources of uranium, economically recoverable at the arbitrary price ceiling of US$130/kg, were enough to last for between 70 and 100 years.

The OECD's red book of 2011 said that conventional uranium resources had grown by 12.5% since 2008 due to increased exploration, with this increase translating into greater than a century of uranium available if the rate of use were to continue at the 2011 level. In 2007, the OECD estimated 670 years of economically recoverable uranium in total conventional resources and phosphate ores assuming the then-current use rate.

Light water reactors make relatively inefficient use of nuclear fuel, mostly fissioning only the very rare uranium-235 isotope. Nuclear reprocessing can make this waste reusable. Newer generation III reactors also achieve a more efficient use of the available resources than the generation II reactors which make up the vast majority of reactors worldwide. With a pure fast reactor fuel cycle with a burn up of all the Uranium and actinides (which presently make up the most hazardous substances in nuclear waste), there is an estimated 160,000 years worth of Uranium in total conventional resources and phosphate ore at the price of 60–100 US$/kg.

Unconventional Fuel Resources

Unconventional uranium resources also exist. Uranium is naturally present in seawater at a concentration of about 3 micrograms per liter, with 4.5 billion tons of uranium considered present in seawater at any time. In 2012 it was estimated that this fuel source could be extracted at 10 times the current price of uranium.

In 2014, with the advances made in the efficiency of seawater uranium extraction, it was suggested that it would be economically competitive to produce fuel for light water reactors from seawater if the process was implemented at large scale. Uranium extracted on an industrial scale from seawater would constantly be replenished by both river erosion of rocks and the natural process of uranium dissolved from the surface area of the ocean floor, both of which maintain the solubility equilibria of seawater concentration at a stable level. Some commentators have argued that this strengthens the case for Nuclear power to be considered a renewable energy.

Breeding

As opposed to light water reactors which use uranium-235 (0.7% of all natural uranium), fast breeder reactors use uranium-238 (99.3% of all natural uranium) or thorium. A number of fuel cycles and breeder reactor combinations are considered to be sustainable and/or renewable sources of energy. In 2006 it was estimated that with seawater extraction, there was likely some five billion years' worth of uranium-238 for use in breeder reactors.

Breeder technology has been used in several reactors, but the high cost of reprocessing fuel safely, at 2006 technological levels, requires uranium prices of more than US$200/kg before becoming justified economically. Breeder reactors are however being pursued as they have the potential to burn up all of the actinides in the present inventory of nuclear waste while also producing power and creating additional quantities of fuel for more reactors via the breeding process.

As of 2017, there are two breeders producing commercial power, BN-600 reactor and the BN-800 reactor, both in Russia. The BN-600, with a capacity of 600 MW, was built in 1980 in Beloyarsk and is planned to produce power until 2025. The BN-800 is an updated version of the BN-600, and started operation in 2014. The Phénix breeder reactor in France was powered down in 2009 after 36 years of operation.

Both China and India are building breeder reactors. The Indian 500 MWe Prototype Fast Breeder Reactor is in the commissioning phase, with plans to build more.

A nuclear fuel rod assembly bundle being inspected before entering a reactor.

Another alternative to fast breeders is thermal breeder reactors that use uranium-233 bred from thorium as fission fuel in the thorium fuel cycle. Thorium is about 3.5 times more common than uranium in the Earth's crust, and has different geographic characteristics. This would extend the total practical fissionable resource base by 450%. India's three-stage nuclear power programme features the use of a thorium fuel cycle in the third stage, as it has abundant thorium reserves but little uranium.

Nuclear Waste

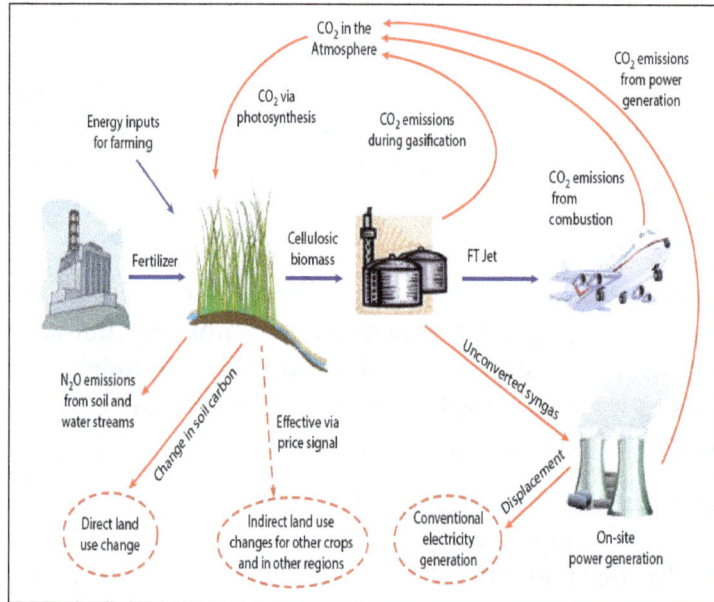

The most important waste stream from nuclear power reactors is spent nuclear fuel. From LWRs, it is typically composed of 95% uranium, 4% fission products from the energy generating nuclear fission reactions, as well as about 1% transuranic actinides (mostly reactor grade plutonium, neptunium and americium) from unavoidable neutron capture events. The plutonium and other transuranics are responsible for the bulk of the long-term radioactivity, whereas the fission products are responsible for the bulk of the short-term radioactivity.

High-level Radioactive Waste

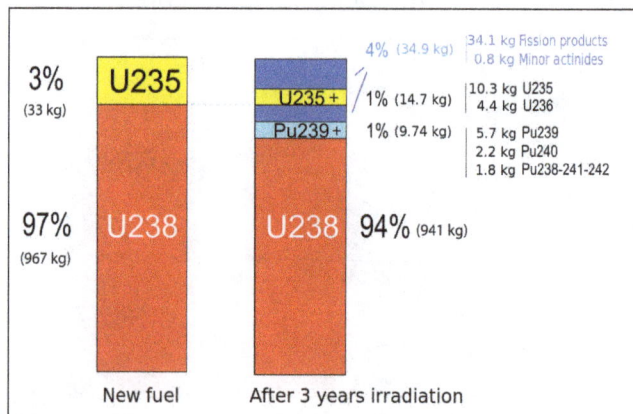

Typical composition of UOx before and after approximately 3 years of fission service in the once-thru fuel cycle of a LWR. Thermal neutron-spectrum-reactors, which presently constitute the majority of the world fleet, cannot burn up the reactor grade plutonium that is generated efficiently, limiting the effective useful fuel life to a few years at most. Reactors in Europe and Asia are permitted to burn later refined MOX fuel, though the burnup is similarly not complete.

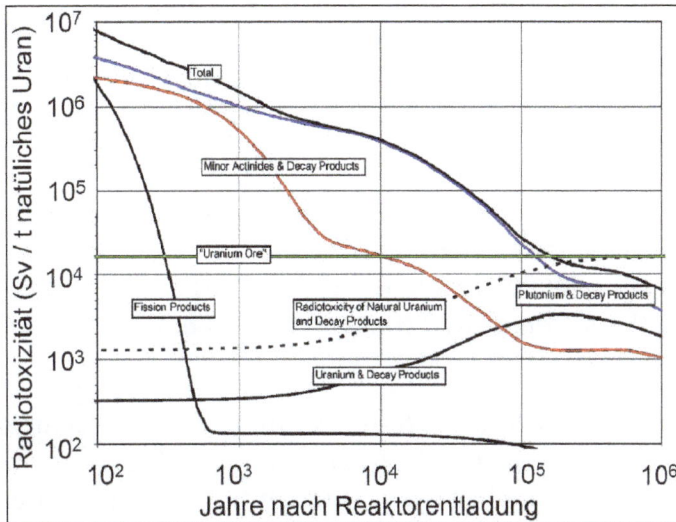

In the years outside a reactor, the activity of spent UOx fuel, in comparison to the activity of natural uranium ore. The various plutonium isotopes that are generated and minor actinides constitute the primary hazard following the relatively rapid decay of the fission products after approximately 300 years. The long lived fission products, Tc-99 and I-129, though less radioactive than the natural uranium ore they derived from, are the focus of much thought on containing.

Following interim storage in a spent fuel pool, the bundles of used fuel rod assemblies of a typical nuclear power station are often stored on site in the likes of the eight dry cask storage vessels pictured above. At Yankee Rowe Nuclear Power Station, which generated 44 billion kilowatt hours of electricity when in service, its complete spent fuel inventory is contained within sixteen casks. It is commonly estimated that to produce a per capita lifetime supply of energy at a western standard of living, approximately 3 GWh, would require on the order of the volume of a soda can of Low enriched uranium per person and thus result in a similar volume of spent fuel generated.

The high-level radioactive waste/spent fuel that is generated from power production, requires treatment, management and isolation from the environment. The technical issues in accomplishing this are considerable, due to the extremely long periods some particularly sublimation prone, mildly radioactive wastes, remain potentially hazardous to living organisms, namely the long-lived fission products, Technetium-99 (half-life 220,000 years) and Iodine-129 (half-life 15.7 million years), which dominate the waste stream in radioactivity after the more intensely radioactive short-lived fission products(SLFPs) have decayed into stable elements, which takes approximately 300 years. To successfully isolate the LLFP waste from the biosphere, either separation and transmutation, or some variation of a synroc treatment and deep geological storage, is commonly suggested.

While in the US, spent fuel is presently in its entirety, federally classified as a nuclear waste and is treated similarly, in other countries it is largely reprocessed to produce a partially recycled fuel, known as mixed oxide fuel or MOX. For spent fuel that does not undergo reprocessing, the most concerning isotopes are the medium-lived transuranic elements, which are led by reactor grade plutonium (half-life 24,000 years).

Some proposed reactor designs, such as the American Integral Fast Reactor and the Molten salt reactor can more completely use or burnup the spent reactor grade plutonium fuel and other minor actinides, generated from light water reactors, as under the designed fast fission spectrum, these elements are more likely to fission and produce the aforementioned fission products in their place. This offers a potentially more attractive alternative to deep geological disposal.

The thorium fuel cycle results in similar fission products, though builds up much less transuranic elements from neutron capture events within a reactor and therefore spent thorium fuel, breeding the true fuel of fissile U-233, is somewhat less concerning from a radiotoxic and security standpoint.

Low-level Radioactive Waste

The nuclear industry also produces a large volume of low-level radioactive waste in the form of contaminated items like clothing, hand tools, water purifier resins, and (upon decommissioning) the materials of which the reactor itself is built. Low-level waste can be stored on-site until radiation levels are low enough to be disposed as ordinary waste, or it can be sent to a low-level waste disposal site.

Waste Relative to other Types

In countries with nuclear power, radioactive wastes account for less than 1% of total industrial toxic wastes, much of which remains hazardous for long periods. Overall, nuclear power produces far less waste material by volume than fossil-fuel based power plants. Coal-burning plants are particularly noted for producing large amounts of toxic and mildly radioactive ash due to concentrating naturally occurring metals and mildly radioactive material in coal. A 2008 report from Oak Ridge National Laboratory concluded that coal power actually results in more radioactivity being released into the environment than nuclear power operation, and that the population effective dose equivalent, or dose to the public from radiation from coal plants is 100 times as much as from the operation of nuclear plants. Although coal ash is much less radioactive than spent nuclear fuel on a weight per weight basis, coal ash is produced in much higher quantities per unit of energy generated, and this is released directly into the environment as fly ash, whereas nuclear plants use shielding to protect the environment from radioactive materials, for example, in dry cask storage vessels.

Waste Disposal

The placement of Nuclear waste flasks, generated during US cold war activities, underground at the WIPP facility. The facility is seen as a potential demonstration, for later civilian generated spent fuel, or constituents of it.

Disposal of nuclear waste is often considered the most politically divisive aspect in the lifecycle of a nuclear power facility. Presently, waste is mainly stored at individual reactor sites and there are over 430 locations around the world where radioactive material continues to accumulate. Some experts suggest that centralized underground repositories which are well-managed, guarded, and monitored, would be a vast improvement. There is an "international consensus on the advisability of storing nuclear waste in deep geological repositories", with the lack of movement of nuclear waste in the 2 billion year old natural nuclear fission reactors in Oklo, Gabon being cited as "a source of essential information today."

Most waste packaging, small-scale experimental fuel recycling chemistry and radiopharmaceutical refinement is conducted within remote-handled Hot cells.

There are no commercial scale purpose built underground high-level waste repositories in operation. However, in Finland the Onkalo spent nuclear fuel repository is under construction as of

2015. The Waste Isolation Pilot Plant (WIPP) in New Mexico has been taking nuclear waste since 1999 from production reactors, but as the name suggests is a research and development facility. In 2014 a radiation leak caused by violations in the use of chemically reactive packaging brought renewed attention to the need for quality control management, along with some initial calls for more R&D into the alternative methods of disposal for radioactive waste and spent fuel. In 2017, the facility was formally reopened after three years of investigation and cleanup, with the resumption of new storage taking place later that year.

Reprocessing

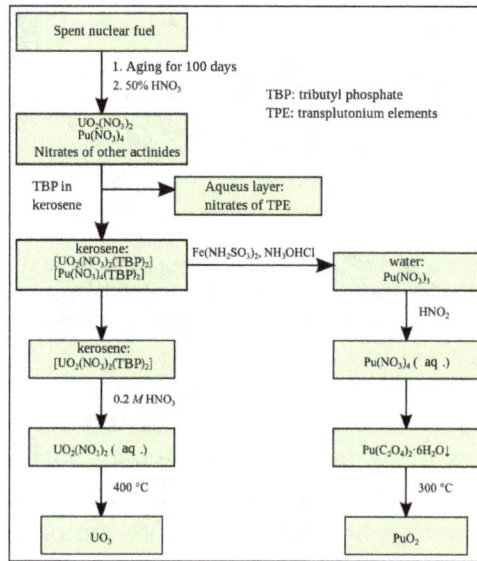

Reprocessing of spent nuclear fuel by the PUREX method, first developed in the 1940s to produce plutonium for nuclear weapons, was demonstrated commercially in Belgium to re-fuel a LWR in the 1960s. This aqueous chemical process continues to be used commercially to separate reactor grade plutonium (RGPu) for reuse as MOX fuel. It remains controversial, as the separated plutonium can be used to make nuclear weapons.

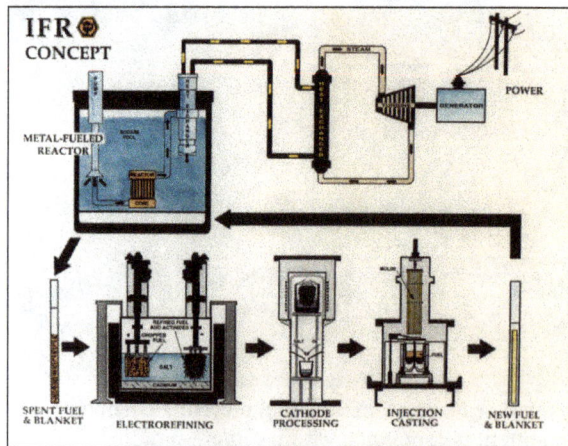

The most developed, though commercially unfielded, alternative reprocessing method, is Pyro-processing, most prominently suggested as part of the metallic-fueled, Integral fast reactor (IFR)

concept proposed in the 1990s. After the spent fuel is dissolved in molten salt, the actinides, consisting mostly of plutonium and uranium, are extracted using electrorefining and/or electrowinning. The resulting mixture of gamma and alpha emitting actinides is mildly self-protecting.

Most thermal reactors run on a once-through fuel cycle, mainly due to the low price of fresh uranium, though many can also fuel made by recycling the fissionable materials in spent nuclear fuel. The most common fissionable material that is recycled is the reactor-grade plutonium (*RGPu*) that is extracted from spent fuel, mixed with uranium oxide and fabricated into mixed-oxide or MOX fuel. The potential for recycling the spent fuel a second time is limited by undesirable neutron economy issues using second-generation MOX fuel in *thermal*-reactors. These issues do not affect fast reactors, which are therefore preferred in order to achieve the full energy potential of the original uranium. The only commercial demonstration of triple burnup to date occurred in the Phénix *fast* reactor.

Because thermal LWRs remain the most common and economically competitive reactor worldwide, the most common form of commercial spent fuel recycling is to recycle the plutonium a single time as MOX fuel, as is done in France, where it is thought to increase the sustainability of the nuclear fuel cycle, reduce the attractiveness of spent fuel to theft and lower the volume of high level nuclear waste. Reprocessing of civilian fuel from power reactors is also currently done in the United Kingdom, Russia, Japan, and India.

The main constituent of spent fuel from the most common light water reactor, is uranium that is slightly more enriched than natural uranium, which can be recycled, though there is a lower incentive to do so. Most of this "recovered uranium", or at times referred to as reprocessed uranium, remains in storage. It can however be used in a fast reactor, used directly as fuel in CANDU reactors, or re-enriched for another cycle through an LWR. The direct use of recovered uranium to fuel a CANDU reactor was first demonstrated at Quishan, China. The first re-enriched uranium reload to fuel a commercial LWR, occurred in 1994 at the Cruas unit 4, France. Re-enriching of reprocessed uranium is common in France and Russia. When reprocessed uranium, namely Uranium-236, is part of the fuel of LWRs, it generates a spent fuel and plutonium isotope stream with greater inherent self-protection, than the once-through fuel cycle.

While reprocessing offers the potential recovery of up to 95% of the remaining uranium and plutonium fuel, in spent nuclear fuel and a reduction in long term radioactivity within the remaining waste. Reprocessing has been politically controversial because of the potential to contribute to nuclear proliferation and varied perceptions of increasing the vulnerability to nuclear terrorism and because of its higher fuel cost, compared to the once-through fuel cycle. Similarly, while reprocessing reduces the volume of high-level waste, it does not reduce the fission products that are the primary residual heat generating and radioactive substances for the first few centuries outside the reactor, thus still requiring an almost identical container-spacing for the initial first few hundred years, within proposed geological waste isolation facilities.

In the United States, spent nuclear fuel is currently not reprocessed. A major recommendation of the Blue Ribbon Commission on America's Nuclear Future was that "the United States should undertake one or more permanent deep geological facilities for the safe disposal of spent fuel and high-level nuclear waste".

The French La Hague reprocessing facility has operated commercially since 1976 and is responsible for half the world's reprocessing as of 2010. Having produced MOX fuel from spent fuel derived from France, Japan, Germany, Belgium, Switzerland, Italy, Spain and the Netherlands, with the non-recyclable part of the spent fuel eventually sent back to the user nation. More than 32,000 tonnes of spent fuel had been reprocessed as of 2015, with the majority from France, 17% from Germany, and 9% from Japan. Once a source of criticism from Greenpeace, more recently the organization have ceased attempting to criticize the facility on technical grounds, having succeeded at performing the process without serious incidents that have been frequent at other such facilities around the world. In the past, the antinuclear movement argued that reprocessing would not be technically or economically feasible.

Nuclear Decommissioning

The financial costs of every nuclear power plant continues for some time after the facility has finished generating its last useful electricity. Once no longer economically viable, nuclear reactors and uranium enrichment facilities are generally decommissioned, returning the facility and its parts to a safe enough level to be entrusted for other uses, such as greenfield status. After a cooling-off period that may last decades, reactor core materials are dismantled and cut into small pieces to be packed in containers for interim storage or transmutation experiments.

In the United States a Nuclear Waste Policy Act and Nuclear Decommissioning Trust Fund is legally required, with utilities banking 0.1 to 0.2 cents/kWh during operations to fund future decommissioning. They must report regularly to the Nuclear Regulatory Commission (NRC) on the status of their decommissioning funds. About 70% of the total estimated cost of decommissioning all U.S. nuclear power reactors has already been collected (on the basis of the average cost of $320 million per reactor-steam turbine unit).

In the United States in 2011, there are 13 reactors that had permanently shut down and are in some phase of decommissioning. With Connecticut Yankee Nuclear Power Plant and Yankee Rowe Nuclear Power Station having completed the process in 2006–2007, after ceasing commercial electricity production circa 1992. The majority of the 15 years, was used to allow the station to naturally cool-down on its own, which makes the manual disassembly process both safer and cheaper. Decommissioning at nuclear sites which have experienced a serious accident are the most expensive and time-consuming.

Installed Capacity and Electricity Production

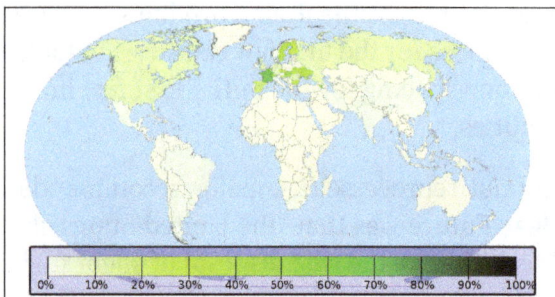

Share of electricity produced by nuclear power in the world.

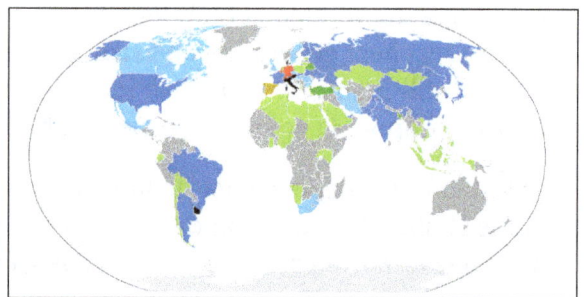

The status of nuclear power globally.

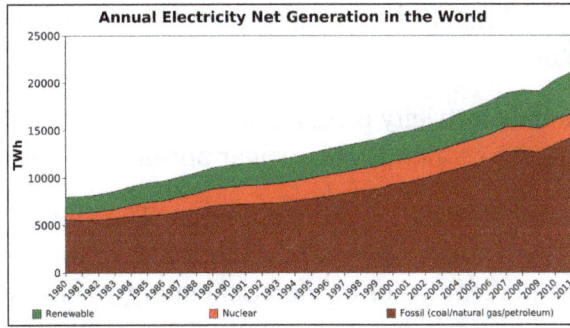

Net electrical generation by source and growth from 1980 to 2010. (Brown) – fossil fuels. (Red) – Fission. (Green) – "all renewables". In terms of energy generated between 1980 and 2010, the contribution from fission grew the fastest.

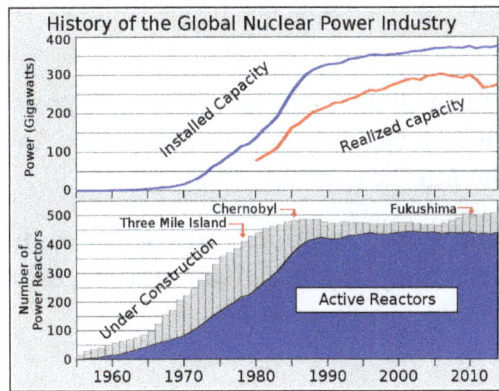

The rate of new construction builds for civilian fission-electric reactors essentially halted in the late 1980s, with the effects of accidents having a chilling effect. Increased capacity factor realizations in existing reactors was primarily responsible for the continuing increase in electrical energy produced during this period. The halting of new builds c. 1985, resulted in greater fossil fuel generation.

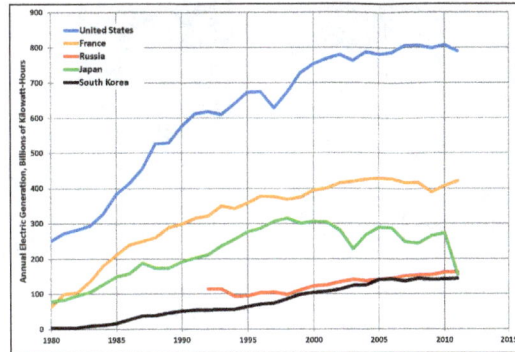

Electricity generation trends in the top five fission-energy producing countries.

Nuclear fission power stations, excluding the contribution from naval nuclear fission reactors, provided 11% of the world's electricity in 2012, somewhat less than that generated by hydro-electric stations at 16%. Since electricity accounts for about 25% of humanity's energy usage with the majority of the rest coming from fossil fuel reliant sectors such as transport, manufacture and home heating, nuclear fission's contribution to the global final energy consumption was about 2.5%. This is a little more than the combined global electricity production from wind, solar, biomass and geothermal power, which together provided 2% of global final energy consumption in 2014.

In addition, there were approximately 140 naval vessels using nuclear propulsion in operation, powered by about 180 reactors.

Nuclear power's share of global electricity production has fallen from 16.5% in 1997 to about 10% in 2017, in large part because the economics of nuclear power have become more difficult.

Regional differences in the use of nuclear power are large. The United States produces the most nuclear energy in the world, with nuclear power providing 20% of the electricity it consumes, while France produces the highest percentage of its electrical energy from nuclear reactors—72% as of 2017. In the European Union as a whole nuclear power provides 30% of the electricity. Nuclear power is the single largest low-carbon electricity source in the United States, and accounts for two-thirds of the European Union's low-carbon electricity. Nuclear energy policy differs among European Union countries, and some, such as Austria, Estonia, Ireland and Italy, have no active nuclear power stations.

Many military and some civilian (such as some icebreakers) ships use nuclear marine propulsion. A few space vehicles have been launched using nuclear reactors: 33 reactors belong to the Soviet RORSAT series and one was the American SNAP-10A.

International research is continuing into additional uses of process heat such as hydrogen production (in support of a hydrogen economy), for desalinating sea water, and for use in district heating systems.

Use in Space

The loading of the Plutonium-238 based MMRTG into the *Mars Curiosity* rover. Assembled in a Hot cell at Idaho National Laboratory.

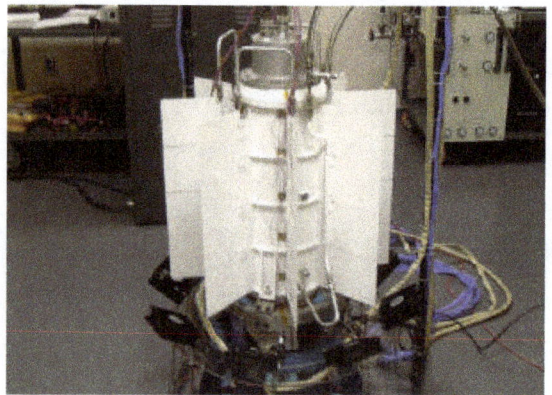

The Multi-mission radioisotope thermoelectric generator (MMRTG), used in several space missions such as the Curiosity Mars rover.

Both fission and fusion appear promising for space propulsion applications, generating higher mission velocities with less reaction mass. This is due to the much higher energy density of nuclear reactions: Some 7 orders of magnitude (10,000,000 times) more energetic than the chemical reactions which power the current generation of rockets.

Radioactive decay has been used on a relatively small scale (few kW), mostly to power space missions and experiments by using radioisotope thermoelectric generators such as those developed at Idaho National Laboratory.

Economics of Nuclear Power Plants

The Ikata Nuclear Power Plant, a pressurized water reactor that cools by utilizing a secondary coolant heat exchanger with a large body of water, an alternative cooling approach to large cooling towers.

The economics of new nuclear power plants is a controversial subject, since there are diverging views on this topic, and multibillion-dollar investments depend on the choice of an energy source. Nuclear power plants typically have high capital costs for building the plant, but low fuel costs. Comparison with other power generation methods is strongly dependent on assumptions about construction timescales and capital financing for nuclear plants as well as the future costs of fossil fuels and renewables as well as for energy storage solutions for intermittent power sources. On the other hand, measures to mitigate global warming, such as a carbon tax or carbon emissions trading, may favor the economics of nuclear power.

Analysis of the economics of nuclear power must also take into account who bears the risks of future uncertainties. To date all operating nuclear power plants have been developed by state-owned or regulated electric utility monopolies Many countries have now liberalized the electricity market where these risks, and the risk of cheaper competitors emerging before capital costs are recovered, are borne by plant suppliers and operators rather than consumers, which leads to a significantly different evaluation of the economics of new nuclear power plants.

Nuclear power plants, though capable of some grid-load following, are typically run as much as possible to keep the cost of the generated electrical energy as low as possible, supplying mostly base-load electricity.

Internationally the price of nuclear plants rose 15% annually in 1970–1990. With PWR stations, having total costs in 2012 of about $96 per megawatt hour (MWh), most of which involves capital construction costs, compared with (in 2018) solar power at $36–44 per MWh, (in 2018) onshore wind at $29–56 per MWH and natural gas at the low end at $64 per MWh. The Fukushima Daiichi nuclear disaster, is expected to increase the costs of operating and new LWR power stations, due to increased requirements for on-site spent fuel management and elevated design basis threats.

Accidents, Attacks and Safety

Nuclear reactors have three unique characteristics that affect their safety, as compared to other power plants. Firstly, intensely radioactive materials are present in a nuclear reactor. Their release to the environment could be hazardous. Secondly, the fission products, which make up most of the intensely radioactive substances in the reactor, continue to generate a significant amount of decay heat even after the fission chain reaction has stopped. If the heat cannot be removed from the reactor, the fuel rods may overheat and release radioactive materials. Thirdly, a rapid increase of the reactor power is possible if the chain reaction cannot be controlled in certain reactor designs. These three characteristics have to be taken into account when designing nuclear reactors.

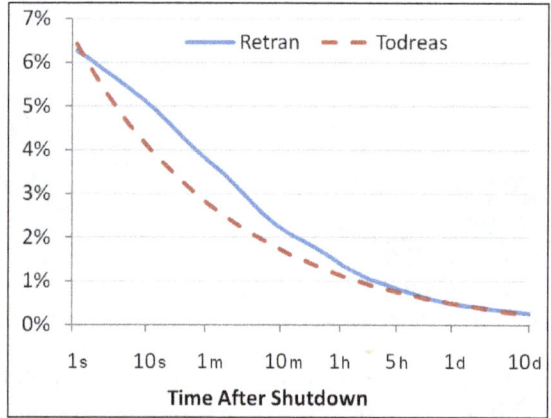

Reactor decay heat as fraction of full power after the reactor shutdown, using two different correlations.
Reactors need cooling after the shutdown of the fission reactions, to prevent a core melt accident.
The loss of cooling caused the Fukushima accident.

Reactors are designed so that an uncontrolled increase of the reactor power is prevented by natural feedback mechanisms: if the temperature or the amount of steam in the reactor increases, the fission power inherently decreases. The chain reaction can be manually stopped by inserting control rods into the reactor core. Emergency core cooling systems can remove the decay heat from the reactor if normal cooling systems fail. Multiple physical barriers limit the release of radioactive materials to the environment even in the case of an accident. The last barrier is the containment.

Accidents

Following the 2011 Fukushima Daiichi nuclear disaster, the world's worst nuclear accident since 1986, 50,000 households were displaced after radiation leaked into the air, soil and sea. Radiation checks led to bans of some shipments of vegetables and fish.

Some serious nuclear and radiation accidents have occurred. The severity of nuclear accidents is generally classified using the International Nuclear Event Scale (INES) introduced by the International Atomic Energy Agency (IAEA). The scale ranks anomalous events or accidents on a scale from 0 (a deviation from normal operation that poses no safety risk) to 7 (a major accident with widespread effects). There have been 3 accidents of level 5 or higher in the civilian nuclear power industry, two of which, the Chernobyl accident and the Fukushima accident, are ranked at level 7.

The Chernobyl accident in 1986 caused approximately 50 deaths from direct and indirect effects, and some temporary serious injuries. The future predicted mortality from cancer increases, is usually estimated at some 4000 in the decades to come. A higher number of the routinely treatable Thyroid cancer, set to be the only type of causal cancer, will likely be seen in future large studies.

The Fukushima Daiichi nuclear accident was caused by the 2011 Tohoku earthquake and tsunami. The accident has not caused any radiation related deaths, but resulted in radioactive contamination of surrounding areas. The difficult Fukushima disaster cleanup will take 40 or more years, and is expected to cost tens of billions of dollars. The Three Mile Island accident in 1979 was a smaller scale accident, rated at INES level 5. There were no direct or indirect deaths caused by the accident.

According to Benjamin K. Sovacool, fission energy accidents ranked first among energy sources in terms of their total economic cost, accounting for 41 percent of all property damage attributed to energy accidents. Another analysis presented in the international journal found that coal, oil, Liquid petroleum gas and hydroelectric accidents (primarily due to the Banqiao dam burst) have resulted in greater economic impacts than nuclear power accidents.

Nuclear power works under an insurance framework that limits or structures accident liabilities in accordance with the Paris convention on nuclear third-party liability, the Brussels supplementary convention, the Vienna convention on civil liability for nuclear damage and the Price-Anderson Act in the United States. It is often argued that this potential shortfall in liability represents an external cost not included in the cost of nuclear electricity; but the cost is small, amounting to about 0.1% of the levelized cost of electricity, according to a CBO study. These beyond-regular-insurance costs for worst-case scenarios are not unique to nuclear power, as hydroelectric power plants are similarly not fully insured against a catastrophic event such as the Banqiao Dam disaster, where 11 million people lost their homes and from 30,000 to 200,000 people died, or large dam failures in general. As private insurers base dam insurance premiums on limited scenarios, major disaster insurance in this sector is likewise provided by the state.

Safety

In terms of lives lost per unit of energy generated, nuclear power has caused fewer accidental deaths per unit of energy generated than all other major sources of energy generation. Energy produced by coal, petroleum, natural gas and hydropower has caused more deaths per unit of energy generated due to air pollution and energy accidents. This is found when comparing the immediate deaths from other energy sources to both the immediate nuclear related deaths from accidents and also including the latent, or predicted, indirect cancer deaths from nuclear energy accidents. When the combined immediate and indirect fatalities from nuclear power and all fossil fuels are compared, including fatalities resulting from the mining of the necessary natural resources to power generation and to air pollution, the use of nuclear power has been calculated to have prevented about 1.8 million deaths between 1971 and 2009, by reducing the proportion of energy that would otherwise have been generated by fossil fuels, and is projected to continue to do so. Following the 2011 Fukushima nuclear disaster, it has been estimated that if Japan had never adopted nuclear power, accidents and pollution from coal or gas plants would have caused more lost years of life.

Forced evacuation from a nuclear accident may lead to social isolation, anxiety, depression, psychosomatic medical problems, reckless behavior, even suicide. Such was the outcome of the 1986 Chernobyl nuclear disaster in Ukraine. A comprehensive 2005 study concluded that "the mental health impact of Chernobyl is the largest public health problem unleashed by the accident to date". Frank N. von Hippel, an American scientist, commented on the 2011 Fukushima nuclear disaster, saying that a disproportionate radiophobia, or "fear of ionizing radiation could have long-term psychological effects on a large portion of the population in the contaminated areas". A 2015 report in *Lancet* explained that serious impacts of nuclear accidents were often not directly attributable to radiation exposure, but rather social and psychological effects. Evacuation and long-term displacement of affected populations created problems for many people, especially the elderly and hospital patients. In January 2015, the number of Fukushima evacuees was around 119,000, compared with a peak of around 164,000 in June 2012.

Attacks and Sabotage

Terrorists could target nuclear power plants in an attempt to release radioactive contamination into the community. The United States 9/11 Commission has said that nuclear power plants were potential targets originally considered for the September 11, 2001 attacks. An attack on a reactor's spent fuel pool could also be serious, as these pools are less protected than the reactor core. The release of radioactivity could lead to thousands of near-term deaths and greater numbers of long-term fatalities.

In the United States, the NRC carries out "Force on Force" (FOF) exercises at all nuclear power plant sites at least once every three years. In the United States, plants are surrounded by a double row of tall fences which are electronically monitored. The plant grounds are patrolled by a sizeable force of armed guards.

Insider sabotage is also a threat because insiders can observe and work around security measures. Successful insider crimes depended on the perpetrators' observation and knowledge of security vulnerabilities. A fire caused 5–10 million dollars worth of damage to New York's Indian Point Energy Center in 1971. The arsonist turned out to be a plant maintenance worker. Some reactors overseas have also reported varying levels of sabotage by workers.

Nuclear Proliferation

Many technologies and materials associated with the creation of a nuclear power program have a dual-use capability, in that they can be used to make nuclear weapons if a country chooses to do so. When this happens a nuclear power program can become a route leading to a nuclear weapon or a public annex to a "secret" weapons program. The concern over Iran's nuclear activities is a case in point.

The Megatons to Megawatts Program was the main driving force behind the sharp reduction in the quantity of nuclear weapons worldwide since the cold war ended. However, without an increase in nuclear reactors and greater demand for fissile fuel, the cost of dismantling has dissuaded Russia from continuing their disarmament.

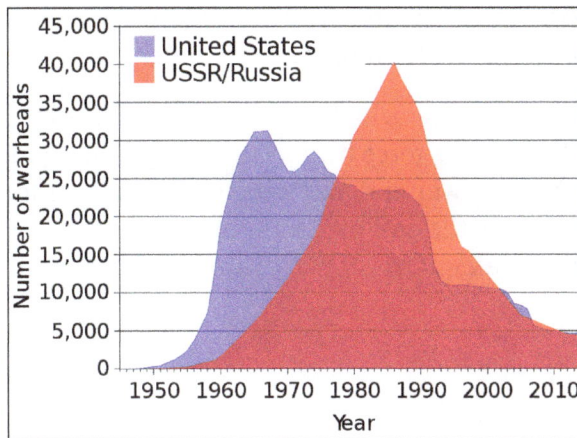

United States and USSR/Russian nuclear weapons stockpiles, 1945–2006.

As of April 2012 there were thirty one countries that have civil nuclear power plants, of which nine have nuclear weapons, with the vast majority of these nuclear weapons states having first produced weapons, before commercial fission electricity stations. Moreover, the re-purposing of civilian nuclear industries for military purposes would be a breach of the Non-proliferation treaty, to which 190 countries adhere.

A fundamental goal for global security is to minimize the nuclear proliferation risks associated with the expansion of nuclear power. The Global Nuclear Energy Partnership was an international effort to create a distribution network in which developing countries in need of energy would receive nuclear fuel at a discounted rate, in exchange for that nation agreeing to forgo their own indigenous develop of a uranium enrichment program. The France-based Eurodif/*European Gaseous Diffusion Uranium Enrichment Consortium* is a program that successfully implemented this concept, with Spain and other countries without enrichment facilities buying a share of the fuel produced at the French controlled enrichment facility, but without a transfer of technology. Iran was an early participant from 1974, and remains a shareholder of Eurodif via Sofidif.

A 2009 United Nations report said that: "The revival of interest in nuclear power could result in the worldwide dissemination of uranium enrichment and spent fuel reprocessing technologies, which present obvious risks of proliferation as these technologies can produce fissile materials that are directly usable in nuclear weapons".

On the other hand, power reactors can also reduce nuclear weapons arsenals when military grade nuclear materials are reprocessed to be used as fuel in nuclear power plants. The Megatons to Megawatts Program, the brainchild of Thomas Neff of MIT, is the single most successful non-proliferation program to date. Up to 2005, the Megatons to Megawatts Program had processed $8 billion of high enriched, weapons grade uranium into low enriched uranium suitable as nuclear fuel for commercial fission reactors by diluting it with natural uranium. This corresponds to the elimination of 10,000 nuclear weapons. For approximately two decades, this material generated nearly 10 percent of all the electricity consumed in the United States (about half of all U.S. nuclear electricity generated) with a total of around 7 trillion kilowatt-hours of electricity produced. Enough energy to energize the entire United States electric grid for about two years. In total it is estimated to have cost $17 billion, a "bargain for US ratepayers", with Russia profiting $12 billion from the deal. Much needed profit for the Russian nuclear oversight industry, which after the collapse of the Soviet economy, had difficulties paying for the maintenance and security of the Russian Federations highly enriched uranium and warheads.

The Megatons to Megawatts Program was hailed as a major success by anti-nuclear weapon advocates as it has largely been the driving force behind the sharp reduction in the quantity of nuclear weapons worldwide since the cold war ended. However without an increase in nuclear reactors and greater demand for fissile fuel, the cost of dismantling and down blending has dissuaded Russia from continuing their disarmament. As of 2013 Russia appears to not be interested in extending the program.

Environmental Impact

Carbon Emissions

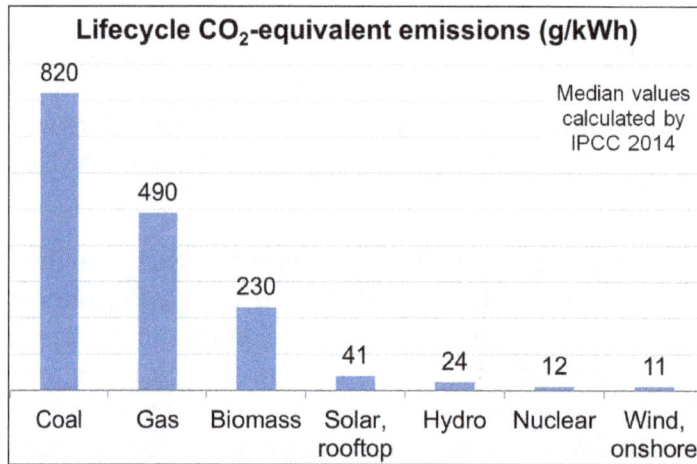

Lifecycle CO_2-equivalent emissions (g/kWh)

820	490	230	41	24	12	11
Coal	Gas	Biomass	Solar, rooftop	Hydro	Nuclear	Wind, onshore

Median values calculated by IPCC 2014

Life-cycle greenhouse gas emissions of electricity supply technologies.

Nuclear power is one of the leading low carbon power generation methods of producing electricity, and in terms of total life-cycle greenhouse gas emissions per unit of energy generated, has emission values comparable to or lower than renewable energy. A 2014 analysis of the carbon footprint literature by the Intergovernmental Panel on Climate Change (IPCC) reported that the embodied total life-cycle emission intensity of fission electricity has a median value of 12g CO_2eq/kWh, which is the lowest out of all commercial baseload energy sources. This is contrasted with coal and natural gas at 820 and 490g CO_2 eq/kWh. From the beginning of its commercialization in the 1970s, nuclear power has prevented the emission of about 64 billion tonnes of carbon dioxide equivalent that would have otherwise resulted from the burning of fossil fuels in thermal power stations.

Radiation

The variation in a person's absorbed natural background radiation, averages 2.4 mSv/a globally but frequently varies between 1 mSv/a and 13 mSv/a depending in most part on the geology a person resides upon. According to the United Nations (UNSCEAR), regular NPP/nuclear power plant operations including the nuclear fuel cycle, increases this amount to 0.0002 millisieverts (mSv) per year of public exposure as a global average. The average dose from operating NPPs to the local populations around them is *less than* 0.0001 mSv/a. The average dose to those living within 50 miles of a coal power plant is over three times this dose, 0.0003 mSv/a.

As of a 2008 report, Chernobyl resulted in the most affected surrounding populations and male recovery personnel receiving an average initial 50 to 100 mSv over a few hours to weeks, while the remaining global legacy of the worst nuclear power plant accident in average exposure is 0.002 mSv/a and is continually dropping at the decaying rate, from the initial high of 0.04 mSv per person averaged over the entire populace of the Northern Hemisphere in the year of the accident in 1986.

Renewable Energy and Nuclear Power

Slowing global warming requires a transition to a low-carbon economy, mainly by burning far less fossil fuel. Limiting global warming to 1.5 degrees C is technically possible if no new fossil fuel power plants are built from 2019. This has generated considerable interest and dispute in determining the best path forward to rapidly replace fossil-based fuels in the global energy mix, with intense academic debate. Sometimes the IEA says that countries without nuclear should develop it as well as their renewable power.

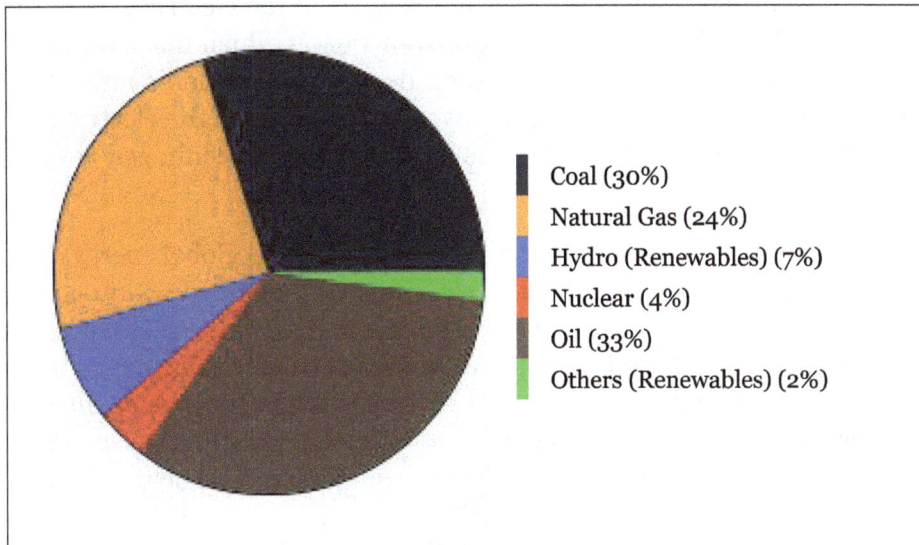

Coal (30%)
Natural Gas (24%)
Hydro (Renewables) (7%)
Nuclear (4%)
Oil (33%)
Others (Renewables) (2%)

World total primary energy consumption, energy for heating, transport, electricity, by source in 2015 was 87% fossil fueled. In the period of 1999 to 2015, this fossil fuel percentage has remained at 87%.

In developed nations the economically feasible geography for new hydropower is lacking, with every geographically suitable area largely already exploited. Proponents of wind and solar energy claim these resources alone could eliminate the need for nuclear power.

Nuclear powered aircraft carriers, presently require as depicted, jet-fuel replenishment at sea operations, by expensive replenishment oilers. The Naval Research Laboratory team led by Heather Willauer has developed a process that is designed to use the ample electrical power onboard carriers for alternative in-situ synthesis of jet-fuel from its chemical building blocks by extracting the carbon dioxide (CO_2) in seawater in tandem with hydrogen (H_2) and recombining the two into long chain hydrocarbon liquids. Willauer estimated that the carbon neutral jet fuel for Navy and Marine aviation, could be synthesized from seawater in quantities up to 100,000 US gal (380,000 L) per day, at a cost of three to six U.S. dollars per gallon. The U.S. Navy is expected to deploy the technology some time in the 2020s.

Some analysts argue that conventional renewable energy sources, wind and solar do not offer the scalability necessary for a large scale decarbonization of the electric grid, mainly due to intermittency-related considerations. Along with other commentators who have questioned the links between the anti-nuclear movement and the fossil fuel industry. These commentators point, in support of the assessment, to the expansion of the coal burning Lippendorf Power Station in Germany and in 2015 the opening of a large, 1730 MW coal burning power station in Moorburg, the only such coal burning facility of its kind to commence operations, in Western Europe in the 2010s. Germany is likely to miss its 2020 emission reduction target.

Several studies suggest that it might be theoretically possible to cover a majority of world energy generation with new renewable sources. The Intergovernmental Panel on Climate Change (IPCC) has said that if governments were supportive, renewable energy supply could account for close to 80% of the world's energy use by 2050.

Analysis in 2015 by professor and chair of Environmental Sustainability Barry W. Brook and his colleagues on the topic of replacing fossil fuels entirely, from the electric grid of the world, has determined that at the historically modest and proven-rate at which nuclear energy was added to and replaced fossil fuels in France and Sweden during each nation's building programs in the 1980s, nuclear energy could displace or remove fossil fuels from the electric grid completely within 10 years, "allowing the world to meet the most stringent greenhouse-gas mitigation targets".

In a similar analysis, Brook had earlier determined that 50% of all global energy, that is not solely electricity, but transportation synthetic fuels etc. could be generated within approximately 30 years, if the global nuclear fission build rate was identical to each of these nation's already proven installation rates in units of installed nameplate capacity, GW per year, per unit of global GDP (GW/year/$). This is in contrast to the conceptual studies for a *100% renewable energy* world, which would require an orders of magnitude more costly global investment per year, which has no historical precedent, along with far greater land that would have to be devoted to the wind, wave and solar projects, and the inherent assumption that humanity will use less, and not more, energy in the future. As Brook notes, the "principal limitations on nuclear fission are not technical, economic or fuel-related, but are instead linked to complex issues of societal acceptance, fiscal and political inertia, and inadequate critical evaluation of the real-world constraints facing [the other] low-carbon alternatives."

In some places which aim to phase out fossil fuels in favor of low carbon power, such as Britain, seasonal energy storage is difficult to provide, so having renewables supply over 60% of electricity might be expensive. As of 2019 whether interconnectors or new nuclear would be more expensive than taking renewables over 60% is still being researched and debated. Britain's older gas-cooled nuclear reactors are not flexible to balance demand, wind and solar, but the island's newer water-cooled reactors should have similar flexibility to fossil fueled power plants. According to the operator from 2025 the British electricity grid may spend periods zero-carbon, with only renewables and nuclear. However actually supplying the electricity grid only from nuclear and renewables may be done together with interconnected countries, such as France in the case of Britain.

Nuclear power is comparable to, and in some cases lower, than many renewable energy sources in terms of lives lost per unit of electricity delivered. However, as opposed to renewable energy, conventional designs for nuclear reactors produce a smaller volume of manufacture and operations related waste, most notably, the intensely radioactive spent fuel that needs to be stored or reprocessed. A nuclear plant also needs to be disassembled and removed and much of the disassembled nuclear plant needs to be stored as low level nuclear waste for a few decades.

In an EU wide 2018 assessment of progress in reducing greenhouse gas emissions per capita, France and Sweden were the only two large industrialized nations within the EU to receive a positive rating, as every other country received a "poor" to "very poor" grade.

A 2018 analysis by MIT argued that, to be much more cost-effective as they approach deep decarbonization, electricity systems should integrate baseload low carbon resources, such as nuclear, with renewables, storage and demand response.

Nuclear power stations require approximately one square kilometer of land per typical reactor. Environmentalists and conservationists have begun to question the global renewable energy expansion proposals, as they are opposed to the frequently controversial use of once forested land to situate renewable energy systems. Seventy five academic conservationists signed a letter, suggesting a more effective policy to mitigate climate change involving the reforestation of this land proposed for renewable energy production, to its prior natural landscape, by means of the native trees that previously inhabited it, in tandem with the lower land use footprint of nuclear energy, as the path to assure both the commitment to carbon emission reductions and to succeed with landscape rewilding programs that are part of the global native species protection and re-introduction initiatives.

These, mostly biological scientists, argue that government commitments to increase renewable energy usage while simultaneously making commitments to expand areas of biological conservation, are two competing land use outcomes, in opposition to one another, that are increasingly coming into conflict. With the existing protected areas for conservation at present regarded as insufficient to safeguard biodiversity "the conflict for space between energy production and habitat will remain one of the key future conservation issues to resolve."

Debate on Nuclear Power

The nuclear power debate concerns the controversy which has surrounded the deployment and use of nuclear fission reactors to generate electricity from nuclear fuel for civilian purposes. The debate about nuclear power peaked during the 1970s and 1980s, when it "reached an intensity unprecedented in the history of technology controversies", in some countries.

Proponents of nuclear energy regard it as a sustainable energy source that reduces carbon emissions and increases energy security by decreasing dependence on imported energy sources. M. King Hubbert, who popularized the concept of peak oil, saw oil as a resource that would run out and considered nuclear energy its replacement. Proponents also claim that the present quantity of nuclear waste is small and can be reduced through the latest technology of newer reactors, and that the operational safety record of fission-electricity is unparalleled.

Opponents believe that nuclear power poses many threats to people and the environment such as the risk of nuclear weapons proliferation and terrorism. They also contend that reactors are complex machines where many things can and have gone wrong. In years past, they also argued that when all the energy-intensive stages of the nuclear fuel chain are considered, from uranium mining to nuclear decommissioning, nuclear power is neither a low-carbon nor an economical electricity source.

Arguments of economics and safety are used by both sides of the debate.

Advanced Fission Reactor Designs

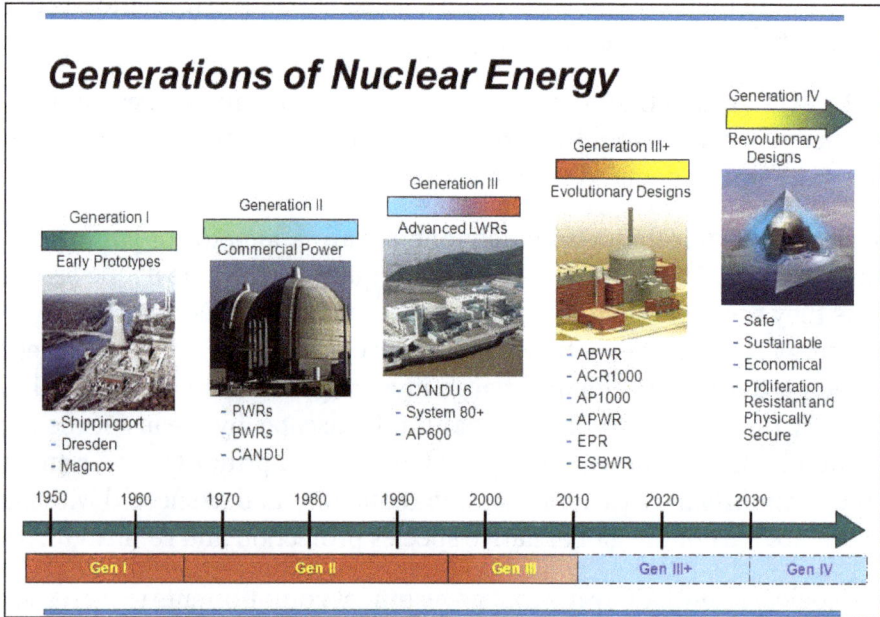

Generation IV roadmap from Argonne National Laboratory.

Current fission reactors in operation around the world are second or third generation systems, with most of the first-generation systems having been already retired. Research into advanced generation IV reactor types was officially started by the Generation IV International Forum (GIF) based on eight technology goals, including to improve economics, safety, proliferation resistance, natural resource utilization and the ability to consume existing nuclear waste in the production of electricity. Most of these reactors differ significantly from current operating light water reactors, and are expected to be available for commercial construction after 2030.

Hybrid Nuclear Fusion-fission

Hybrid nuclear power is a proposed means of generating power by use of a combination of nuclear fusion and fission processes. The concept dates to the 1950s, and was briefly advocated by Hans Bethe during the 1970s, but largely remained unexplored until a revival of interest in 2009, due to delays in the realization of pure fusion. When a sustained nuclear fusion power plant is built, it has the potential to be capable of extracting all the fission energy that remains in spent fission fuel, reducing the volume of nuclear waste by orders of magnitude, and more importantly, eliminating all actinides present in the spent fuel, substances which cause security concerns.

Nuclear Fusion

Nuclear fusion reactions have the potential to be safer and generate less radioactive waste than fission. These reactions appear potentially viable, though technically quite difficult and have yet to be created on a scale that could be used in a functional power plant. Fusion power has been under theoretical and experimental investigation since the 1950s.

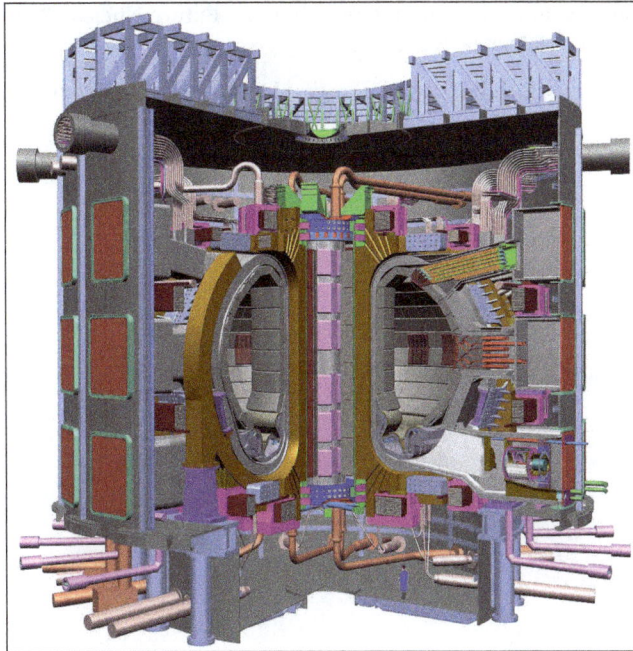

Schematic of the ITER tokamak under construction.

Several experimental nuclear fusion reactors and facilities exist. The largest and most ambitious international nuclear fusion project currently in progress is ITER, a large tokamak under construction in France. ITER is planned to pave the way for commercial fusion power by demonstrating self-sustained nuclear fusion reactions with positive energy gain. Construction of the ITER facility began in 2007, but the project has run into many delays and budget overruns. The facility is now not expected to begin operations until the year 2027–11 years after initially anticipated. A follow on commercial nuclear fusion power station, DEMO, has been proposed. There are also suggestions for a power plant based upon a different fusion approach, that of an inertial fusion power plant.

Fusion powered electricity generation was initially believed to be readily achievable, as fission-electric power had been. However, the extreme requirements for continuous reactions and plasma containment led to projections being extended by several decades. In 2010, more than 60 years after the first attempts, commercial power production was still believed to be unlikely before 2050.

References

- Renewable-energy, encyclopedia: nationalgeographic.org, Retrieved 1 August, 2019

- Terry S. Reynolds, Stronger than a Hundred Men: A History of the Vertical Water Wheel, JHU Press, 2002 ISBN 0-8018-7248-0

- Geothermal-energy, science: britannica.com, Retrieved 5 March, 2019

- Fossil-fuel, science: britannica.com, Retrieved 3 May, 2019

- "Korea's nuclear phase-out policy takes shape". World Nuclear News. 19 June 2017. Retrieved 12 February 2018

- Petroleum, , science: britannica.com, Retrieved 5 February, 2019

- "World Nuclear Power Reactors | Uranium Requirements | Future Nuclear Power". Www.world-nuclear.org. World Nuclear Association. Retrieved 8 May 2018

- Non-renewable-energy, encyclopedia: nationalgeographic.org, Retrieved 8 June, 2019

Energy Consumption and Efficiency

3

- Energy Consumption
- World Energy Consumption
- Resource Depletion
- Energy Security
- Energy Policy
- Energy Crisis
- Energy Efficiency
- Energy Monitoring and Targeting
- Energy Efficiency in Transport

The total amount of energy or power used is termed as energy consumption. Energy efficiency refers to the practices which are aimed at reducing the wastage of energy required to provide products and services. The chapter closely examines the key concepts of energy consumption and efficiency such as energy monitoring and energy crisis to provide an extensive understanding of the subject.

Energy Consumption

We human beings have been using vast proportions of earth's natural resources for our own needs. We use energy for or heating and cooling, lighting, heating water and operating appliances. Apart from that we use energy for many purposes, such as traveling in airplanes and cars using oil that is converted into gasoline.

The U.S. department of energy has divided energy users into 3 category: Residential and Commercial,

Industrial and Transportation. Residential and commercial uses energy to light up their homes, for heating and cooling purposes and to fulfill their daily basic needs. Industrial sector mainly consume energy for lighting up of offices, running machines, for heating and cooling purposes. Transportation sector uses energy for uploading and downloading of goods and services from one place to another. Their mainly source of energy is oil on which transportation sector depends.

According to U.S. Energy Information Administration (EIA), the demand for global energy is is projected to grow 44% between 2005 and 2030, driven by robust economic growth and expanding populations in the world's developing countries. It has also been reported that the dependence on coal has increased sharply by the developing countries in the last few years and will continue to increase unless these nations change their existing laws and strategies and particularly those related to greenhouse gas emissions, robust growth in coal use is likely to continue.

These projections are driven by strong long-term economic growth in the world's developing nations. The current global economic downturn will dampen world energy demand in the near term, as manufacturing and consumer demand for goods and services slows; however, with economic recovery anticipated to begin within the next 12 to 24 months, most nations are expected to see energy consumption growth at rates anticipated prior to the recession.

The report also states that china leads in the usage of coal and thus the consumption of coal in the country has doubled since 2000. Given the country's rapidly expanding economy and large domestic coal deposits, its demand for coal is projected to remain strong. In the reference case, coal use is projected to expand by 2% every year between 2005 and 2030, and coal's share of total world energy consumption is expected to reach 29% in 2030.

Both China and India will be the key energy consumers in the future. Both the countries were consuming an average of 10% of world's total energy consumption in 1990 but in 2006 their combined share was 19 percent. Strong economic growth in both countries continues over the projection period, with their combined energy use increasing nearly twofold and making up 28 percent of world energy consumption in 2030.

The report also finds concerns that with the increase in the prices of fossil fuels, energy security and greenhouse emissions will drive the country towards the development of nuclear generating capacity. World nuclear capacity is all set to grow between 374GW in 2005 to 498GW in 2030. China is projected to add 45 GW of net nuclear capacity over the projection period. Russia is expected to add 18 GW, and India is at its heels, with 17 GW. By 2030, the U.S. will have added 15 GW of nuclear power, says the EIA.

World Energy Consumption

World energy consumption is the total energy produced and used by the entire human civilization. Typically measured per year, it involves all energy harnessed from every energy source applied towards humanity's endeavors across every single industrial and technological sector, across every country. It does not include energy from food, and the extent to which direct biomass burning has been accounted for is poorly documented. Being the power source metric of civilization, world energy consumption has deep implications for humanity's socio-economic-political sphere.

Institutions such as the International Energy Agency (IEA), the U.S. Energy Information Administration (EIA), and the European Environment Agency (EEA) record and publish energy data periodically. Improved data and understanding of world energy consumption may reveal systemic trends and patterns, which could help frame current energy issues and encourage movement towards collectively useful solutions.

Closely related to energy consumption is the concept of total primary energy supply (TPES), which – on a global level – is the sum of energy production minus storage changes. Since changes of energy storage over the year are minor, TPES values can be used as an estimator for energy consumption. However, TPES ignores conversion efficiency, overstating forms of energy with poor conversion efficiency (e.g. coal, gas and nuclear) and understating forms already accounted for in converted forms (e.g. photovoltaics or hydroelectricity). The IEA estimates that, in 2013, total primary energy supply (TPES) was 1.575×10^{17} Wh (= 157.5 PWh, 157,500 TWh, 5.67×10^{20} joules, or 13,541 Mtoe) or about 18 TW-year. From 2000–2012 coal was the source of energy with the total largest growth. The use of oil and natural gas also had considerable growth, followed by hydropower and renewable energy. Renewable energy grew at a rate faster than any other time in history during this period. The demand for nuclear energy decreased, in part due to nuclear disasters (Three Mile Island in 1979, Chernobyl in 1986, and Fukushima in 2011). More recently, consumption of coal has declined relative to renewable energy. Coal dropped from about 29% of the global total primary energy consumption in 2015 to 27% in 2017, and non-hydro renewables were up to about 4% from 2%.

In 2011, expenditures on energy totaled over US$6 trillion, or about 10% of the world gross domestic product (GDP). Europe spends close to one-quarter of the world's energy expenditures, North America close to 20%, and Japan 6%.

Energy Supply, Consumption and Electricity

Key figures (TWh)		
Year	Primary energy supply (TPES)[1]	Final energy consumption[1]	Electricity generation
1973	71,013 (Mtoe 6,106)	54,335 (Mtoe 4,672)	6,129
1990	102,569	–	11,821
2000	117,687	–	15,395
2010	147,899 (Mtoe 12,717)	100,914 (Mtoe 8,677)	21,431
2011	152,504 (Mtoe 13,113)	103,716 (Mtoe 8,918)	22,126
2012	155,505 (Mtoe 13,371)	104,426 (Mtoe 8,979)	22,668
2013	157,482 (Mtoe 13,541)	108,171 (Mtoe 9,301)	23,322
2014	155,481 (Mtoe 13,369)	109,613 (Mtoe 9,425)	23,816
2015	158,715 (Mtoe 13,647)	109,136 (Mtoe 9,384)	

[1]converted from Mtoe into TWh (1 Mtoe = 11.63 TWh) and from Quad BTU into TWh (1 Quad BTU = 293.07 TWh)

World total primary energy supply (TPES), or "primary energy" differs from the world final energy consumption because much of the energy that is acquired by humans is lost as other forms of energy during the process of its refinement into usable forms of energy and its transport from its initial place of supply to consumers. For instance, when oil is extracted from the ground it must be refined into gasoline, so that it can be used in a car, and transported over long distances to gas stations where it can be used by consumers. World final energy consumption refers to the fraction of the world's primary energy that is used in its final form by humanity.

Also one needs to bear in mind that there are different qualities of energy. Heat, especially at a relatively low temperature, is low-quality energy, whereas electricity is high-quality energy. It takes around 3 kWh of heat to produce 1 kWh of electricity. But by the same token, a kilowatt-hour of this high-quality electricity can be used to pump several kilowatt-hours of heat into a building using a heat pump. And electricity can be used in many ways in which heat cannot. So the "loss" of energy incurred when generating electricity is not the same as a loss due, say, to resistance in power lines.

In 2014, world primary energy supply amounted to 155,481 terawatt-hour (TWh) or 13,541 Mtoe, while the world final energy consumption was 109,613 TWh or about 29.5% less than the total supply. World final energy consumption includes products as lubricants, asphalt and petrochemicals which have chemical energy content but are not used as fuel. This non-energy use amounted to 9,723 TWh (836 Mtoe) in 2015.

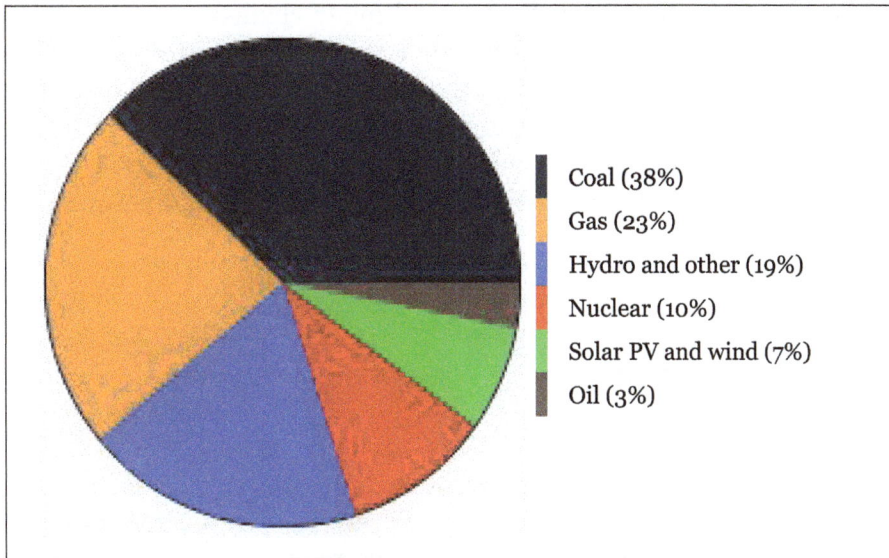

2018 World electricity generation (26,700 TWh).

The United States Energy Information Administration (EIA) regularly publishes a report on world consumption for most types of primary energy resources. For 2013, estimated world energy consumption was 5.67×10^{20} joules, or 157,481 TWh. According to the IEA the total world energy consumption in past years was 143,851 TWh in 2008, 133,602 TWh in 2005, 117,687 TWh in 2000, and 102,569 TWh in 1990. In 2012 approximately 22% of world energy was consumed in North America, 5% was consumed South and Central America, 23% was consumed in Europe and Eurasia, 3% was consumed in Africa, and 40% was consumed in the Asia Pacific region.

Electricity Generation

The total amount of electricity consumed worldwide was 19,504 TWh in 2013, 16,503 TWh in 2008, 15,105 TWh in 2005, and 12,116 TWh in 2000. By the end of 2014, the total installed electricity generating capacity worldwide was nearly 6.142 TW (million MW) which only includes generation connected to local electricity grids. In addition there is an unknown amount of heat and electricity consumed off-grid by isolated villages and industries. In 2014, the share of world energy consumption for electricity generation by source was coal at 41%, natural gas at 22%, nuclear at 11%, hydro at 16%, other sources (solar, wind, geothermal, biomass, etc.) at 6% and oil at 4%. Coal and natural gas were the most used energy fuels for generating electricity. The world's electricity consumption was 18,608 TWh in 2012. This figure is about 18% smaller than the generated electricity, due to grid losses, storage losses, and self-consumption from power plants (gross generation). Cogeneration (CHP) power stations use some of the heat that is otherwise wasted for use in buildings or in industrial processes.

In 2016 while total world energy came from 80% fossil fuels, 10% biofuels, 5% nuclear and 5% renewable (hydro, wind, solar, geothermal), only 18% of that total world energy was in the form of electricity. Most of the other 82% was used for heat and transportation.

Recently there has been a large increase in international agreements and national Energy Action Plans, such as the EU 2009 Renewable Energy Directive, to increase the use of renewable energy due to the growing concerns about pollution from energy sources that come from fossil fuels such as oil, coal, and natural gas. One such initiative was the United Nations Development Programme's

World Energy Assessment in 2000 that highlighted many challenges humanity would have to overcome in order to shift from fossil fuels to renewable energy sources. From 2000–2012 renewable energy grew at a rate higher than any other point in history, with a consumption increase of 176.5 million tonnes of oil. During this period, oil, coal, and natural gas continued to grow and had increases that were much higher than the increase in renewable energy. The following figures illustrate the growth in consumption of fossil fuels such as oil, coal, and natural gas as well as renewable sources of energy during this period.

Trends

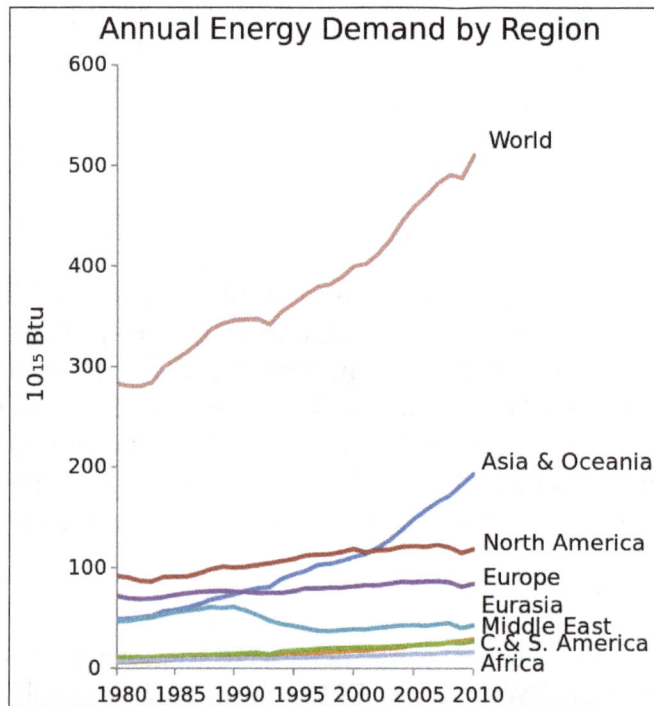

World primary energy consumption in quadrillion Btu.

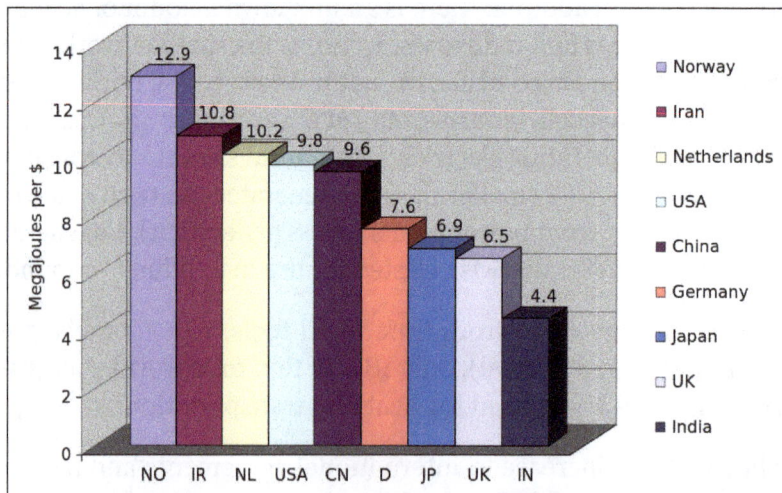

Energy intensity of different economies: The graph shows the ratio between energy usage and GDP for selected countries. GDP is based on 2004 purchasing power parity and 2000 dollars adjusted for inflation.

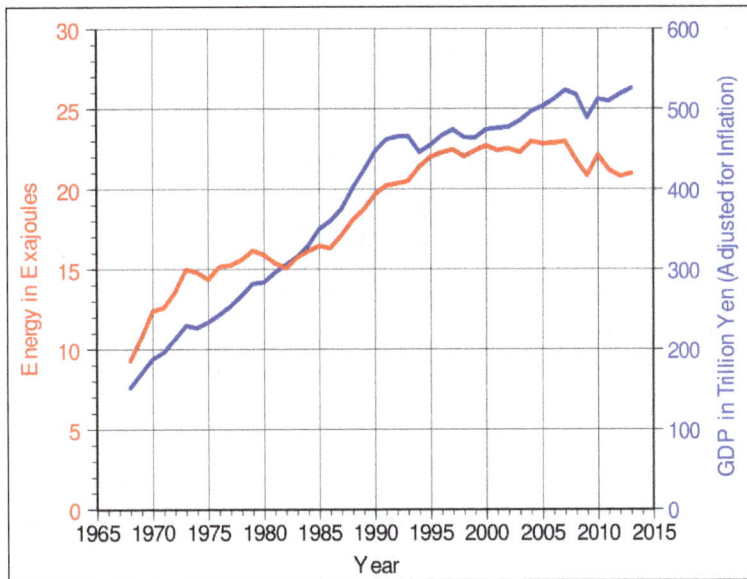

GDP and energy consumption in Japan, 1958–2000

The data shows the correlation between GDP and energy use; however, it also shows that this link can be broken. After the oil shocks of 1973 and 1979 the energy use stagnated while Japan's GDP continued to grow, after 1985, under the influence of the then much cheaper oil, energy use resumed its historical relation to GDP.

The energy consumption growth in the G20 slowed down to 2% in 2011, after the strong increase of 2010. The economic crisis is largely responsible for this slow growth. For several years now, the world energy demand is characterized by the bullish Chinese and Indian markets, while developed countries struggle with stagnant economies, high oil prices, resulting in stable or decreasing energy consumption.

According to IEA data from 1990 to 2008, the average energy use per person increased 10% while world population increased 27%. Regional energy use also grew from 1990 to 2008: the Middle East increased by 170%, China by 146%, India by 91%, Africa by 70%, Latin America by 66%, the US by 20%, the EU-27 block by 7%, and world overall grew by 39%.

In 2008, total worldwide primary energy consumption was 132,000 terawatt-hours (TWh) or 474 exajoules (EJ). In 2012, primary energy demand increased to 158,000 TWh (567 EJ).

Energy consumption in the G20 increased by more than 5% in 2010 after a slight decline of 2009. In 2009, world energy consumption decreased for the first time in 30 years by 1.1%, or about 130 million tonnes of oil equivalent (Mtoe), as a result of the financial and economic crisis, which reduced world GDP by 0.6% in 2009.

This evolution is the result of two contrasting trends: Energy consumption growth remained vigorous in several developing countries, specifically in Asia (+4%). Conversely, in OECD, consumption was severely cut by 4.7% in 2009 and was thus almost down to its 2000 levels. In North America, Europe and the CIS, consumptions shrank by 4.5%, 5% and 8.5% respectively due to the slowdown in economic activity. China became the world's largest energy consumer (18% of the total) since its consumption surged by 8% during 2009 (up from 4% in 2008). Oil remained the largest energy source (33%) despite the fact that its share has been decreasing over time. Coal posted a growing role in the world's energy consumption: in 2009, it accounted for 27% of the total.

Most energy is used in the country of origin, since it is cheaper to transport final products than raw materials. In 2008, the share export of the total energy production by fuel was: oil 50% (1,952/3,941 Mt), gas 25% (800/3,149 bcm) and hard coal 14% (793/5,845 Mt).

Most of the world's high energy resources are from the conversion of the sun's rays to other energy forms after being incident upon the planet. Some of that energy has been preserved as fossil energy, some is directly or indirectly usable; for example, via solar PV/thermal, wind, hydro- or wave power. The total solar irradiance is measured by satellite to be roughly 1361 watts per square meter, though it fluctuates by about 6.9% during the year due to the Earth's varying distance from the sun. This value, after multiplication by the cross-sectional area intercepted by the Earth, is the total rate of solar energy received by the planet; about half, 89,000 TW, reaches the Earth's surface.

The estimates of remaining non-renewable worldwide energy resources vary, with the remaining fossil fuels totaling an estimated 0.4 yottajoule (YJ) or 4×10^{23} joules, and the available nuclear fuel such as uranium exceeding 2.5 YJ. Fossil fuels range from 0.6 to 3 YJ if estimates of reserves of methane clathrates are accurate and become technically extractable. The total power flux from the sun intercepting the Earth is 5.5 YJ per year, though not all of this is available for human consumption. The IEA estimates for the world to meet global energy demand for the two decades from 2015 to 2035 it will require investment of $48 trillion and "credible policy frameworks."

According to IEA the goal of limiting warming to 2 °C is becoming more difficult and costly with each year that passes. If action is not taken before 2017, CO_2 emissions would be locked-in by energy infrastructure existing in 2017. Fossil fuels are dominant in the global energy mix, supported by $523 billion subsidies in 2011, up almost 30% on 2010 and six times more than subsidies to renewables.

Regional energy use (kWh/capita & TWh) and growth 1990–2008 (%)								
	kWh/capita			Population (million)			Energy use (1,000 TWh)		
Region	1990	2008	Growth	1990	2008	Growth	1990	2008	Growth
US	89,021	87,216	−2%	250	305	22%	22.3	26.6	20%
EU-28	40,240	40,821	1%	473	499	5%	19.0	20.4	7%
Middle East	19,422	34,774	79%	132	199	51%	2.6	6.9	170%
China	8,839	18,608	111%	1,141	1,333	17%	10.1	24.8	146%
Latin America	11,281	14,421	28%	355	462	30%	4.0	6.7	66%
Africa	7,094	7,792	10%	634	984	55%	4.5	7.7	70%
India	4,419	6,280	42%	850	1,140	34%	3.8	7.2	91%
Others	25,217	23,871	nd	1,430	1,766	23%	36.1	42.2	17%
The World	19,422	21,283	10%	5,265	6,688	27%	102.3	142.3	39%

Emissions

Global warming emissions resulting from energy production are an environmental problem. Efforts to resolve this include the Kyoto Protocol, 1997 and the Paris Agreement, 2015, international governmental agreements aiming to reduce harmful climate impacts, which a number of nations have signed. Limiting global temperature increase to 2 degrees Celsius, thought to be a risk by the SEI, is now doubtful.

To limit global temperature to a hypothetical 2 degrees Celsius rise would demand a 75% decline in carbon emissions in industrial countries by 2050, if the population is 10 billion in 2050. Across 40 years, this averages to a 2% decrease every year. In 2011, the emissions of energy production continued rising regardless of the consensus of the basic problem. Hypothetically, according to Robert Engelman (Worldwatch Institute), in order to prevent collapse, human civilization would have to stop increasing emissions within a decade regardless of the economy or population.

Greenhouse gases are not the only emissions of energy production and consumption. Large amounts of pollutants such as sulphurous oxides (SO_x), nitrous oxides (NO_x), and particulate matter (PM) are produced from the combustion of fossil fuels and biomass; the World Health Organization estimates that 7 million premature deaths are caused each year by air pollution. Biomass combustion is a major contributor. In addition to producing air pollution like fossil fuel combustion, most biomass has high CO_2 emissions.

By Source

Fossil Fuels

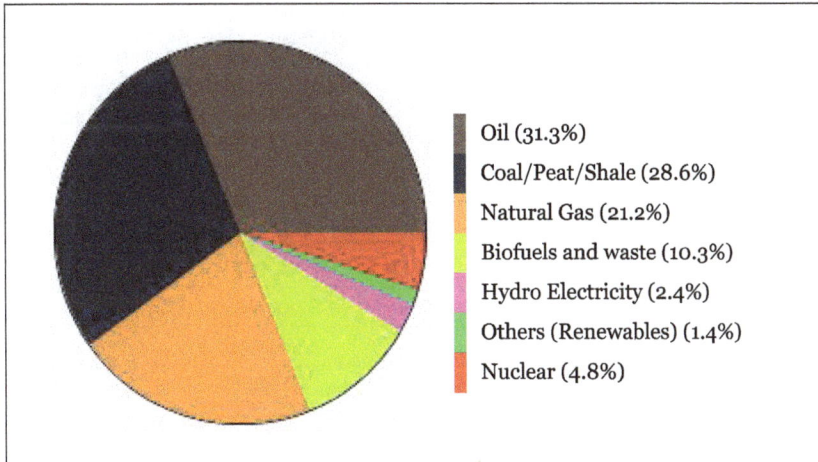

Total primary energy supply of 13,699 mega-toe.

The twentieth century saw a rapid twenty-fold increase in the use of fossil fuels. Between 1980 and 2006, the worldwide annual growth rate was 2%. According to the US Energy Information Administration's 2006 estimate, the estimated 471.8 EJ total consumption in 2004, was divided as given in the table, with fossil fuels supplying 86% of the world's energy.

Coal

In 2000, China accounted for 28% of world coal consumption, other Asia consumed 19%, North

America 25% and the EU 14%. The single greatest coal-consuming country is China. Its share of the world coal production was 28% in 2000 and rose to 48% in 2009. In contrast to China's ~70% increase in coal consumption, world coal use increased 48% from 2000 to 2009. In practice, the majority of this growth occurred in China and the rest in other Asia. China's energy consumption is mostly driven by the industry sector, the majority of which comes from coal consumption.

World annual coal production increased 1,905 Mt or 32% in 6 years in 2011 compared to 2005, of which over 70% was in China and 8% in India. Coal production was in 2011 7,783 Mt, and 2009 6,903 Mt, equal to 12.7% production increase in two years.

If production and consumption of coal continue at the rate as in 2008, proven and economically recoverable world reserves of coal would last for about 150 years. This is much more than needed for an irreversible climate catastrophe. Coal is the largest source of carbon dioxide emissions in the world. According to James Hansen the single most important action needed to tackle the climate crisis is to reduce CO_2 emissions from coal. Indonesia and Australia exported together 57.1% of the world coal export in 2011. China, Japan, South Korea, India and Taiwan had 65% share of all the world coal import in 2011.

Regional coal supply (TWh), share 2010 (%) and share of change 2000–2010						
Region	2000	2008	2009	2010	%	Change 2000–2009
North America	6,654	6,740	6,375	6,470	16%	−1.2%
Asia excl. China	5,013	7,485	7,370	7,806	19%	18.9%
China	7,318	16,437	18,449	19,928	48%	85.5%
EU	3,700	3,499	3,135	3,137	8%	−3.8%
Africa	1,049	1,213	1,288	1,109	3%	0.4%
Russia	1,387	1,359	994	1,091	3%	−2.0%
Others	1,485	1,763	1,727	1,812	4%	2.2%
Total	26,607	38,497	39,340	41,354	100%	47.9%

Top 10 coal exporters (Mt)					
Rank	Nation	2010	2011	Share % 2011	2012
1	Indonesia	162	309	29.7%	383
2	Australia	298	285	27.4%	302
3	Russia	89	99	9.5%	103
4	US	57	85	8.2%	106
5	Colombia	68	76	7.3%	82
6	South Africa	68	70	6.7%	72
7	Kazakhstan	33	34	3.3%	32
8	Canada	24	24	2.3%	25
9	Vietnam	21	23	2.2%	18
10	Mongolia	17	22	2.1%	22
x	Others	19	14	1.3%	
Total (Mt)		856	1,041		1,168
Top ten		97.8%	98.7%		

Oil

Coal fueled the industrial revolution in the 18th and 19th century. With the advent of the automobile, aeroplanes and the spreading use of electricity, oil became the dominant fuel during the twentieth century. The growth of oil as the largest fossil fuel was further enabled by steadily dropping prices from 1920 until 1973. After the oil shocks of 1973 and 1979, during which the price of oil increased from 5 to 45 US dollars per barrel, there was a shift away from oil. Coal, natural gas, and nuclear became the fuels of choice for electricity generation and conservation measures increased energy efficiency. In the U.S. the average car more than doubled the number of miles per gallon. Japan, which bore the brunt of the oil shocks, made spectacular improvements and now has the highest energy efficiency in the world. From 1965 to 2008, the use of fossil fuels has continued to grow and their share of the energy supply has increased. From 2003 to 2008, coal was the fastest growing fossil fuel.

It is estimated that between 100 and 135 billion tonnes of oil has been consumed between 1850 and the present.

Natural Gas

In 2009, the world use of natural gas grew 31% compared to 2000. 66% of this growth was outside EU, North America, Latin America, and Russia. Others include the Middle East, Asia, and Africa. The gas supply increased also in the previous regions: 8.6% in the EU and 16% in the North America 2000–2009.

Regional gas supply (TWh) and share 2010 (%)					
Land	2000	2008	2009	2010	%
North America	7,621	7,779	8,839	8,925	27%
Asia excl. China	2,744	4,074	4,348	4,799	14%
China	270	825	1,015	1,141	3%
EU	4,574	5,107	4,967	5,155	16%
Africa	612	974	1,455	1,099	3%
Russia	3,709	4,259	4,209	4,335	13%
Latin America	1,008	1,357	958	nd	nd
Others	3,774	5,745	6,047	7,785	23%
Total	24,312	30,134	31,837	33,240	100%

Nuclear Power

As of 1 July 2016, the world had 444 operable grid-electric nuclear fission power reactors with 62 others under construction.

Annual generation of nuclear power has been on a slight downward trend since 2007, decreasing 1.8% in 2009 to 2558 TWh, and another 1.6% in 2011 to 2518 TWh, despite increases in production from most countries worldwide, because those increases were more than offset by decreases in Germany and Japan. Nuclear power met 11.7% of the world's electricity demand in 2011.

While all the commercial reactors today use nuclear fission energy, there are plans to use nuclear fusion energy for future power plants. Several international nuclear fusion reactor experiments exists or are being constructed, including ITER.

Renewable Energy

Renewable energy is generally defined as energy that comes from resources that are not significantly depleted by their use, such as sunlight, wind, rain, tides, waves and geothermal heat. Renewable energy is gradually replacing conventional fuels in four distinct areas: electricity generation, hot water/space heating, motor fuels, and rural (off-grid) energy services.

Based on REN21's 2014 report, renewables contributed 19 percent to our energy consumption and 22 percent to our electricity generation in 2012 and 2013, respectively. This energy consumption is divided as 9% coming from traditional biomass, 4.2% as heat energy (non-biomass), 3.8% hydro electricity and 2% electricity from wind, solar, geothermal, and biomass. Worldwide investments in renewable technologies amounted to more than US$214 billion in 2013, with countries like China and the United States heavily investing in wind, hydro, solar and biofuels. Renewable energy resources exist over wide geographical areas, in contrast to other energy sources, which are concentrated in a limited number of countries. Rapid deployment of renewable energy and energy efficiency is resulting in significant energy security, climate change mitigation, and economic benefits. In international public opinion surveys there is strong support for promoting renewable sources such as solar power and wind power. At the national level, at least 30 nations around the world already have renewable energy contributing more than 20 percent of energy supply. National renewable energy markets are projected to continue to grow strongly in the coming decade and beyond.

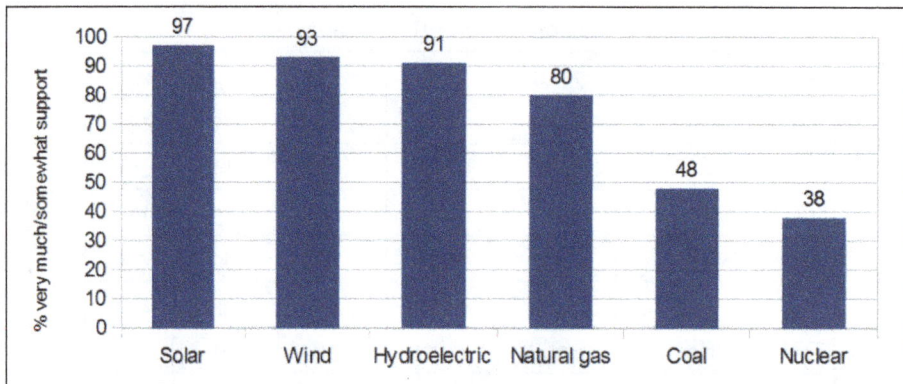

Strong public support for renewables worldwide in 2011

The following table shows increasing nameplate capacity, and has capacity factors that range from 11% for solar, to 40% for hydropower.

Selected renewable energy global indicators	2008	2009	2010	2011	2012	2013	2014	2015
Investment in new renewable capacity (annual) (10⁹ USD)	182	178	237	279	256	232	270	285
Renewables power capacity (existing) (GWe)	1,140	1,230	1,320	1,360	1,470	1,578	1,712	1,849
Hydropower capacity (existing) (GWe)	885	915	945	970	990	1,018	1,055	1,064
Wind power capacity (existing) (GWe)	121	159	198	238	283	319	370	433

Solar PV capacity (grid-connected) (GWe)	16	23	40	70	100	138	177	227
Solar hot water capacity (existing) (GWth)	130	160	185	232	255	373	406	435
Ethanol production (annual) (10^9 litres)	67	76	86	86	83	87	94	98
Biodiesel production (annual) (10^9 litres)	12	17.8	18.5	21.4	22.5	26	29.7	30
Countries with policy targets for renewable energy use	79	89	98	118	138	144	164	173

Renewable energy 2000-2013 (TWh)			
	2000	2010	2013
North-America	1,973	2,237	2,443
EU	1,204	2,093	2,428
Russia	245	239	271
China	2,613	3,374	3,847
Asia (-China)	4,147	4,996	5,361
Africa	2,966	3,930	4,304
Latin America	1 502	2,127	2,242
Other	567	670	738
Total renewable	15,237	19,711	21,685
Total energy	116,958	148,736	157,485
Share	13.0%	13.3%	13.8%
Total nonrenewable	101,721	129,025	135,800

From 2000 to 2013 the total renewable energy use has increased by 6,450 TWh and total energy use by 40,500 TWh.

Hydro

Hydroelectricity is the term referring to electricity generated by hydropower; the production of electrical power through the use of the kinetic energy of falling or flowing water. In 2015 hydropower generated 16.6% of the world's total electricity and 70% of all renewable electricity, which continues the rapid rate of increase experienced between 2003 and 2009. Hydropower is produced in 150 countries, with the Asia-Pacific region generating 32 percent of global hydropower in 2010. China is the largest hydroelectricity producer, with 2,600 PJ (721 TWh) of production in 2010, representing around 17% of domestic electricity use. There are now three hydroelectricity plants larger than 10 GW: the Three Gorges Dam in China, Itaipu Dam in Brazil, and Guri Dam in Venezuela. Nine of the worlds top 10 renewable electricity producers are primarily hydroelectric, one is wind.

Marine Energy

Marine energy, also known as *ocean energy* and *marine and hydrokinetic energy* (MHK) includes tidal and wave power and is a relatively new sector of renewable energy, with most projects still in the pilot phase, but the theoretical potential is equivalent to 4–18 million tonne of oil equivalent (toe). MHK development in U.S. and international waters includes projects using devices such as, wave energy converters in open coastal areas with significant waves, tidal turbines placed in coastal and estuarine areas, in-stream turbines in fast-moving rivers, ocean current turbines in areas of strong marine currents, and ocean thermal energy converters in deep tropical waters.

Wind

Wind power is growing at the rate of 17% annually, with a worldwide installed capacity of 432,883 megawatts (MW) at the end of 2015, and is widely used in Europe, Asia, and the United States. Several countries have achieved relatively high levels of wind power penetration, such as 21% of stationary electricity production in Denmark, 18% in Portugal, 16% in Spain, 14% in Ireland and 9% in Germany in 2010. As of 2011, 83 countries around the world are using wind power on a commercial basis. Continuing strong growth, by 2016 wind generated 3% of global power annually.

Solar

Solar energy, radiant light and heat from the sun, has been harnessed by humans since ancient times using a range of ever-evolving technologies. Solar energy technologies include solar heating, solar photovoltaics, concentrated solar power and solar architecture, which can make considerable contributions to solving some of the most urgent problems the world now faces. The International Energy Agency projected that solar power could provide "a third of the global final energy demand after 2060, while CO_2 emissions would be reduced to very low levels." Solar technologies are broadly characterized as either passive solar or active solar depending on the way they capture, convert and distribute solar energy. Active solar techniques include the use of photovoltaic systems and solar thermal collectors to harness the energy. Passive solar techniques include orienting a building to the Sun, selecting materials with favorable thermal mass or light dispersing properties, and designing spaces that naturally circulate air. From 2012 to 2016 solar capacity tripled and now provides 1.3% of global energy.

Geothermal

Geothermal energy is used commercially in over 70 countries. In 2004, 200 petajoules (56 TWh) of electricity was generated from geothermal resources, and an additional 270 petajoules (75 TWh) of geothermal energy was used directly, mostly for space heating. In 2007, the world had a global capacity for 10 GW of electricity generation and an additional 28 GW of direct heating, including extraction by geothermal heat pumps. Heat pumps are small and widely distributed, so estimates of their total capacity are uncertain and range up to 100 GW.

Bioenergy

Until the beginning of the nineteenth century biomass was the predominant fuel, today it has only a small share of the overall energy supply. Electricity produced from biomass sources was estimated at 44 GW for 2005. Biomass electricity generation increased by over 100% in Germany, Hungary, the Netherlands, Poland, and Spain. A further 220 GW was used for heating (in 2004), bringing the total energy consumed from biomass to around 264 GW. The use of biomass fires for cooking is excluded. World production of bioethanol increased by 8% in 2005 to reach 33 gigalitres (8.7×10^9 US gal), with most of the increase in the United States, bringing it level to the levels of consumption in Brazil. Biodiesel increased by 85% to 3.9 gigalitres (1.0×10^9 US gal), making it the fastest growing renewable energy source in 2005. Over 50% is produced in Germany.

By Sector

World energy use by sector, 2012			
Sector	10^{15}Btu	Petawatt-hours	%
Residential	53.0	15.5	13
Commercial	29.3	8.6	7
Industrial	222.3	65.1	54
Transportation	104.2	30.5	26
Total	408.9	119.8	100

The table shows the amounts of energy consumed worldwide in 2012 by four sectors:

- Residential (heating, lighting, and appliances).

- Commercial (lighting, heating and cooling of commercial buildings, and provision of water and sewer services).

- Industrial users (agriculture, mining, manufacturing, and construction).

- Transportation (passenger, freight, and pipeline).

Of the total 120 PWh (120×10^{15} Wh) consumed, 19.4 were in the form of electricity, but this electricity required 61.7 PWh to produce. Thus the total energy consumption was around 160 PWh (ca 550×10^{15} Btu). The efficiency of a typical existing power plant is around 38%. The new generation of gas-fired plants reaches a substantially higher efficiency of 55%. Coal is the most common fuel for the world's electricity plants.

Another report gives different values for the sectors, apparently due to different definitions. According to this, total world energy use per sector in 2008 was industry 28%, transport 27% and residential and service 36%. Division was about the same in the year 2000.

World energy use per sector				
Year	2000	2008	2000	2008
Sector	TWh		%	
Industry	21,733	27,273	27	28
Transport	22,563	26,742	28	27
Residential and service	30,555	35,319	37	36
Non-energy use	7,119	8,688	9	9
Total	81,970	98,022	100	100

European Union

The European Environmental Agency (EEA) measures final energy consumption (does not include energy used in production and lost in transportation) and finds that the transport sector is responsible for 32% of final energy consumption, households 26%, industry 26%, services 14% and agriculture 3% in 2012. The use of energy is responsible for the majority of greenhouse gas emissions (79%), with the energy sector representing 31%, transport 19%, industry 13%, households 9% and others 7%.

While efficient energy use and resource efficiency are growing as public policy issues, more than 70% of coal plants in the European Union are more than 20 years old and operate at an efficiency level of between 32–40%. Technological developments in the 1990s have allowed efficiencies in the range of 40–45% at newer plants. However, according to an impact assessment by the European Commission, this is still below the best available technological (BAT) efficiency levels of 46–49%. With gas-fired power plants the average efficiency is 52% compared to 58–59% with best available technology (BAT), and gas and oil boiler plants operate at average 36% efficiency (BAT delivers 47%). According to that same impact assessment by the European Commission, raising the efficiency of all new plants and the majority of existing plants, through the setting of authorisation and permit conditions, to an average generation efficiency of 52% in 2020 would lead to a reduction in annual consumption of 15 km³ (3.6 cu mi) of natural gas and 25 Mt (25,000,000 long tons; 28,000,000 short tons) of coal.

Resource Depletion

Resource depletion is the consumption of a resource faster than it can be replenished. Natural resources are commonly divided between renewable resources and non-renewable resources. Use of either of these forms of resources beyond their rate of replacement is considered to be resource depletion. The value of a resource is a direct result of its availability in nature and the cost of extracting the resource, the more a resource is depleted the more the value of the resource increases. There are several types of resource depletion the most known being; Aquifer depletion, deforestation, mining for fossil fuels and minerals, pollution or contamination of resources, slash-and-burn agricultural practices, Soil erosion, and overconsumption, excessive or unnecessary use of resources.

Resource depletion is most commonly used in reference to farming, fishing, mining, water usage, and consumption of fossil fuels. Depletion of wildlife populations is called *defaunation*.

Depletion Accounting

In an effort to offset the depletion of resources theorists have come up with depletion accounting or better known as 'green accounting'. Depletion accounting aims to account for nature's value on an equal footing with the market economy. Resource depletion accounting uses data provided from countries to estimate the adjustments needed due to their use and depletion of the natural capital available to them. Natural capital are natural resources such as mineral deposits or timber stocks. Depletion accounting factors in several different influences such as the number of years until resource exhaustion, the cost of resource extraction and the demand of the resource. Resource extraction industries make up a large part of the economic activity in developing countries, this in turn leads to higher levels of resource depletion and environmental degradation in developing countries. Theorists argue that implementation of resource depletion accounting is necessary in developing countries. Depletion accounting also seeks to measure the social value of natural resources and ecosystems. Measurement of social value is sought through ecosystem services which are defined as the benefits of nature to households, communities and economies.

Importance

There are many different groups interested in depletion accounting. Environmentalists are interested in depletion accounting as a way to track the use of natural resources over time, hold governments accountable or to compare their environmental conditions to those of another country. Economists want to measure resource depletion to understand how financially reliant countries or corporations are on non-renewable resources, whether this use can be sustained and the financial drawbacks of switching to renewable resources in light of the depleting resources.

Issues

Depletion accounting is complex to implement as nature is not as quantifiable like cars, houses or bread. For depletion accounting to work, appropriate units of natural resources must be established so that natural resources can be viable in the market economy. The main issues that arise when trying to do so are, determining a suitable unit of account, deciding how to deal with "collective" nature of a complete ecosystem, delineating the borderline of the ecosystem and defining the extent of possible duplication when the resource interacts in more than one ecosystem. Some economists want to include measurement of the benefits arising from public goods provided by nature, but currently there are no market indicators of value. Globally, environmental economics has not been able to provide a consensus of measurement units of nature's services.

Minerals Depletion

Minerals are needed to provide food, clothing, and housing. A United States Geological Survey (USGS) study found a significant long-term trend over the 20th century for non-renewable resources such as minerals to supply a greater proportion of the raw material inputs to the non-fuel, non-food sector of the economy; an example is the greater consumption of crushed stone, sand, and gravel used in construction.

Large-scale exploitation of minerals began in the Industrial Revolution around 1760 in England and has grown rapidly ever since. Technological improvements have allowed humans to dig deeper and access lower grades and different types of ore over that time. Virtually all basic industrial metals (copper, iron, bauxite, etc.), as well as rare earth minerals, face production output limitations from time to time, because supply involves large up-front investments and is therefore slow to respond to rapid increases in demand.

Minerals projected by some to enter production decline:

- Gasoline.

- Copper. Data from the United States Geological Survey (USGS) suggest that it is very unlikely that copper production will peak before 2040.

- Zinc. Developments in hydrometallurgy have transformed non-sulfide zinc deposits (largely ignored until now) into large low cost reserves.

Minerals projected by some to enter production decline during the present century:

- Aluminium.

- Coal.

- Iron.

Oil

Peak oil is the period when the maximum rate of global petroleum extraction is reached, after which the rate of production will undergo a long-term decline. The 2005 Hirsch report concluded that the decreased supply combined with increasing demand will significantly increase the world-wide prices of petroleum derived products, and that most significant will be the availability and price of liquid fuel for transportation.

The Hirsch report, funded by United States Department of Energy, concluded that "The peaking of world oil production presents the U. S. and the world with an unprecedented risk management problem. As peaking is approached, liquid fuel prices and price volatility will increase dramatically, and, without timely mitigation, the economic, social, and political costs will be unprecedented. Viable mitigation options exist on both the supply and demand sides, but to have substantial impact, they must be initiated more than a decade in advance of peaking."

Deforestation

Deforestation is the clearing of forests by cutting or burning of trees and plants in a forested area. As a result of deforestation, presently about one half of the forests that once covered Earth have been destroyed. It occurs for many different reasons, and it has several negative implications on the atmosphere and the quality of the land in and surrounding the forest.

Deforestation is the removal of a forest or stand of trees from land, the wood is harvested as a resource for production of consumer products and firewood for heat. The land then either left to recover and then will be replanted or is converted to non-forest land used as agricultural land or development of urban areas.

Causes

One of the main causes of deforestation is clearing forests for agricultural reasons. As the population of developing areas, especially near rainforests, increases, the need for land for farming becomes more and more important. For most people, a forest has no value when its resources are not being used, so the incentives to deforest these areas outweigh the incentives to preserve the forests. For this reason, the economic value of the forests is very important for the developing countries.

Environmental Impact

Because deforestation is so extensive, it has made several significant impacts on the environment, including carbon dioxide in the atmosphere, changing the water cycle, an increase in soil erosion, and a decrease in biodiversity. Deforestation is often cited as a contributor to global warming. Because trees and plants remove carbon dioxide and emit oxygen into the atmosphere, the reduction of forests contribute to about 12% of anthropogenic carbon dioxide emissions. One of the most pressing issues that deforestation creates is soil erosion. The removal of trees causes higher rates of erosion, increasing risks of landslides, which is a direct threat to many people living close to deforested areas. As forests get destroyed, so does the habitat for millions of animals. It is estimated that 80% of the world's known biodiversity lives in the rainforests, and the destruction of these rainforests is accelerating extinction at an alarming rate.

Controlling Deforestation

The United Nations and the World Bank created programs such as Reducing Emissions from Deforestation and Forest Degradation (REDD), which works especially with developing countries to use subsidies or other incentives to encourage citizens to use the forest in a more sustainable way. In addition to making sure that emissions from deforestation are kept to a minimum, an effort to educate people on sustainability and helping them to focus on the long-term risks is key to the success of these programs. The New York Declaration on Forests and its associated actions promotes reforestation, which is being encouraged in many countries in an attempt to repair the damage that deforestation has done.

Wetlands

Wetlands are ecosystems that are often saturated by enough surface or groundwater to sustain vegetation that is usually adapted to saturated soil conditions, such as cattails, bulrushes, red maples, wild rice, blackberries, cranberries, and peat moss. Because some varieties of wetlands are rich in minerals and nutrients and provide many of the advantages of both land and water environments they contain diverse species and provide a distinct basis for the food chain. Wetland habitats contribute to environmental health and biodiversity. Wetlands are a nonrenewable resource on a human timescale and in some environments cannot ever be renewed. Recent studies indicate that global loss of wetlands could be as high as 87% since 1700 AD, with 64% of wetland loss occurring since 1900. Some loss of wetlands resulted from natural causes such as erosion, sedimentation, subsidence, and a rise in the sea level.

Wetlands provide environmental services for:

- Food and habitat.

- Improving water quality.

- Commercial fishing.

- Floodwater reduction.

- Shoreline stabilization.

- Recreation.

Resources in Wetlands

Some of the world's most successful agricultural areas are wetlands which have been drained an converted to farmland for large-scale agriculture. Large-scale draining of wetlands also occurs for real estate development and urbanization. In contrast in some cases wetlands are also flooded to be converted to recreational lakes or hydro-power generation. In some countries ranchers have also moved their property onto wetlands for grazing due to the nutrient rich vegetation. Wetlands in Southern America also prove a fruitful resource for poachers, as animals with valuable hides such a jaguars, maned wolves, caimans and snakes are drawn to wetlands. The effect of the removal of large predators is still unknown in South African wetlands.

Humans benefit from wetlands in indirect ways as well. Wetlands act as natural water filters, when runoff from either natural or man-made processes pass through, wetlands can have a neutralizing effect. If a wetland is in between an agricultural zone and a freshwater ecosystem, fertilizer runoff will be absorbed by the wetland and used to fuel the slow processes that occur happen, by the time the water reaches the freshwater ecosystem there won't be enough fertilizer to cause destructive algal blooms that poison freshwater ecosystems.

Bramiana Wetlands.

Non-natural causes of wetland degradation:

- Hydrologic alteration:

  ○ Drainage,

  ○ Dredging,

  ○ Stream channelization,

- ○   Ditching,

- ○   Levees,

- ○   Deposition of fill material,

- ○   Stream diversion,

- ○   Groundwater drainage,

- ○   Impoundment.

- •   Urbanization and urban development.

- •   Marinas/boats.

- •   Industrialization and industrial development.

- •   Agriculture.

- •   Silviculture/Timber harvest.

- •   Mining.

- •   Atmospheric deposition.

To preserve the resources extracted from wetlands, current strategies are to rank wetlands and prioritize the conservation of wetlands with more environmental services, create more efficient irrigation for wetlands being used for agriculture and restricting access to wetlands by tourists.

Groundwater

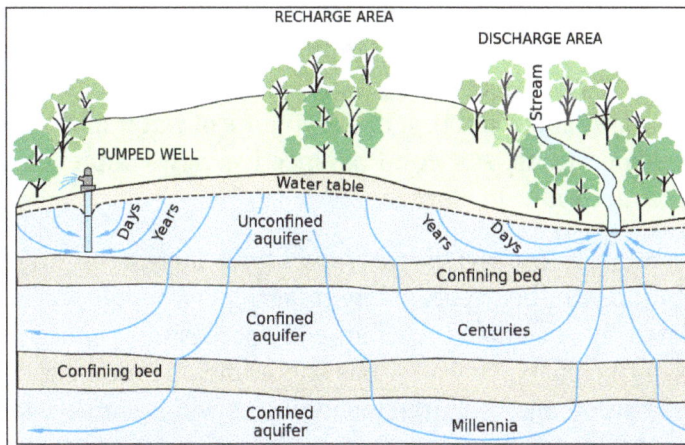

Groundwater flow paths vary greatly in length, depth and travel time from points of recharge to points of discharge in the groundwater system.

Water is an essential resource needed to survive everyday life. Historically, water has had a profound influence on a nation's prosperity and success around the world. Groundwater is water that is in saturated zones underground, the upper surface of the saturated zone is called the water table. Groundwater is held in the pores and fractures of underground materials like sand, gravel and other rock, these rock materials are called aquifers. Groundwater can either flow naturally out of rock materials or can be pumped out. Groundwater supplies wells and aquifers for private, agricultural, and public use and is used by more than a third of the world's population every day for their drinking water. Globally there is 22.6 million cubic kilometers of groundwater available and only .35 million of that is renewable.

Groundwater as a Non-renewable Resource

Groundwater is considered to be a non-renewable resource because less than six percent of the water around the world is replenished and renewed on a human timescale of 50 years. People are already using non-renewable water that is thousands of years old, in areas like Egypt they are using water that may have been renewed a million years ago which is not renewable on human timescales. Of the groundwater used for agriculture 16 to 33% is non-renewable. It is estimated that since the 1960's groundwater extraction has more than doubled, which has increased groundwater depletion. Due to this increase in depletion, in some of the most depleted areas use of groundwater for irrigation has become impossible or cost prohibitive.

Environmental Impacts

Overusing groundwater, old or young can lower subsurface water levels and dry up streams, which could have a huge effect on ecosystems on the surface. When the most easily recoverable fresh groundwater is removed this leaves a residual with inferior water quality. This is in part from induced leakage from the land surface, confining layers or adjacent aquifers that contain saline or contaminated water. Worldwide the magnitude of groundwater depletion from storage may be so large as to constitute a measurable contributor to sea-level rise.

Mitigation

Currently, societies respond to water-resource depletion by shifting management objectives from location and developing new supplies to augmenting conserving and reallocation of existing supplies. There are two different perspectives to groundwater depletion, the first is that depletion is considered literally and simply as a reduction in the volume of water in the saturated zone, regardless of water quality considerations. A second perspective views depletion as a reduction in the usable volume of fresh groundwater in storage.

Augmenting supplies can mean improving water quality or increasing water quantity. Depletion due to quality considerations can be overcome by treatment, whereas large volume metric depletion can only be alleviated by decreasing discharge or increasing recharge. Artificial recharge of storm flow and treated municipal wastewater, has successfully reversed groundwater declines. In the future improved infiltration and recharge technologies will be more widely used to maximize the capture of runoff and treated wastewater.

Renewable Resources

Renewable energy can be collected from renewable resources. The two main sources of renewable energy are solar energy and wind power. The government and scientists are researching and looking upon alternatives to replace the depleting nonrenewable resources. Japan and the U.S. are leading in the department of selling and manufacturing solar powered utilities.

Energy Security

Energy security is the association between national security and the availability of natural resources for energy consumption. Access to (relatively) cheap energy has become essential to the functioning of modern economies. However, the uneven distribution of energy supplies among countries has led to significant vulnerabilities. International energy relations have contributed to the globalization of the world leading to energy security and energy vulnerability at the same time.

In the context of energy security, security of energy supply is an issue of utmost importance. Moreover, it is time to define "a global energy policy model, which not only aims at ensuring an efficient environmental protection but also at ensuring security of energy supply".

Renewable resources and significant opportunities for energy efficiency exist over wide geographical areas, in contrast to other energy sources, which are concentrated in a limited number of countries. Rapid deployment of renewable energy and energy efficiency, and technological diversification of energy sources, would result in significant energy security and economic benefits.

Threats

The modern world relies on a vast energy supply to fuel everything from transportation to communication, to security and health delivery systems. Perhaps most alarmingly, peak oil expert Michael Ruppert has claimed that for every calorie of food produced in the industrial world, ten calories of oil and gas energy are invested in the forms of fertilizer, pesticide, packaging, transportation, and running farm equipment. Energy plays an important role in the national security of any given country as a fuel to power the economic engine. Some sectors rely on energy more heavily than others; for example, the Department of Defense relies on petroleum for approximately 77% of its energy needs. Threats to energy security include the political instability of several energy producing countries, the manipulation of energy supplies, the competition over energy sources, attacks on supply infrastructure, as well as accidents, natural disasters, terrorism, and reliance on foreign countries for oil.

Foreign oil supplies are vulnerable to unnatural disruptions from in-state conflict, exporters' interests, and non-state actors targeting the supply and transportation of oil resources. The political and economic instability caused by war or other factors such as strike action can also prevent the proper functioning of the energy industry in a supplier country. For example, the nationalization of oil in Venezuela has triggered strikes and protests in which Venezuela's oil production rates have yet to recover. Exporters may have political or economic incentive to limit their foreign sales or cause disruptions in the supply chain. Since Venezuela's nationalization of oil, anti-American Hugo Chávez threatened to cut off supplies to the United States more than once. The 1973 oil embargo against the United States is a historical example in which oil supplies were cut off to the United States due to U.S. support of Israel during the Yom Kippur War. This has been done to apply pressure during economic negotiations—such as during the 2007 Russia–Belarus energy dispute. Terrorist attacks targeting oil facilities, pipelines, tankers, refineries, and oil fields are so common they are referred to as "industry risks". Infrastructure for producing the resource is extremely vulnerable to sabotage. One of the worst risks to oil transportation is the exposure of the five ocean chokepoints, like the Iranian-controlled Strait of Hormuz. Anthony Cordesman, a scholar at the Center for Strategic and International Studies in Washington, D.C., warns, "It may take only one asymmetric or conventional attack on a Ghawar Saudi oil field or tankers in the Strait of Hormuz to throw the market into a spiral."

Long-term Security

Long-term measures to increase energy security center on reducing dependence on any one source of imported energy, increasing the number of suppliers, exploiting native fossil fuel or renewable energy resources, and reducing overall demand through energy conservation measures. It can also involve entering into international agreements to underpin international energy trading relationships, such as the Energy Charter Treaty in Europe. All the concern coming from security threats on oil sources long term security measures will help reduce the future cost of importing and exporting fuel into and out of countries without having to worry about harm coming to the goods being transported.

The impact of the 1973 oil crisis and the emergence of the OPEC cartel was a particular milestone that prompted some countries to increase their energy security. Japan, almost totally dependent on imported oil, steadily introduced the use of natural gas, nuclear power, high-speed mass transit systems, and implemented energy conservation measures. The United Kingdom began exploiting North Sea oil and gas reserves, and became a net exporter of energy into the 2000s.

In other countries energy security has historically been a lower priority. The United States, for example, has continued to increase its dependency on imported oil although, following the oil price increases since 2003, the development of biofuels has been suggested as a means of addressing this.

Increasing energy security is also one of the reasons behind a block on the development of natural gas imports in Sweden. Greater investment in native renewable energy technologies and energy conservation is envisaged instead. India is carrying out a major hunt for domestic oil to decrease its dependency on OPEC, while Iceland is well advanced in its plans to become energy independent by 2050 through deploying 100% renewable energy.

Short-term Security

Petroleum

Petroleum, otherwise known as "crude oil", has become the resource most used by countries all around the world including Russia, China (actually, China is mostly dependent on coal (70.5% in 2010)) and the United States of America. With all the oil wells located around the world energy security has become a main issue to ensure the safety of the petroleum that is being harvested. In the middle east oil fields become main targets for sabotage because of how heavily countries rely on oil. Many countries hold strategic petroleum reserves as a buffer against the economic and political impacts of an energy crisis. All 28 members of the International Energy Agency hold a minimum of 90 days of their oil imports, for example.

The value of such reserves was demonstrated by the relative lack of disruption caused by the 2007 Russia-Belarus energy dispute, when Russia indirectly cut exports to several countries in the European Union.

Due to the theories in peak oil and need to curb demand, the United States military and Department of Defense had made significant cuts, and have been making a number of attempts to come up with more efficient ways to use oil.

Natural Gas

Compared to petroleum, reliance on imported natural gas creates significant short-term vulnerabilities. The gas conflicts between Ukraine and Russia of 2006 and 2009 serve as vivid examples of this. Many European countries saw an immediate drop in supply when Russian gas supplies were halted during the Russia-Ukraine gas dispute in 2006.

Natural gas has been a viable source of energy in the world. Consisting of mostly methane, natural gas is produced using two methods: biogenic and thermogenic. Biogenic gas comes from methanogenic organisms located in marshes and landfills, whereas thermogenic gas comes from the anaerobic decay of organic matter deep under the Earth's surface. Russia is the current leading country in production of natural gas.

One of the biggest problems currently facing natural gas providers is the ability to store and transport it. With its low density, it is difficult to build enough pipelines in North America to transport sufficient natural gas to match demand. These pipelines are reaching near capacity and even at full capacity do not produce the amount of gas needed.

Nuclear Power

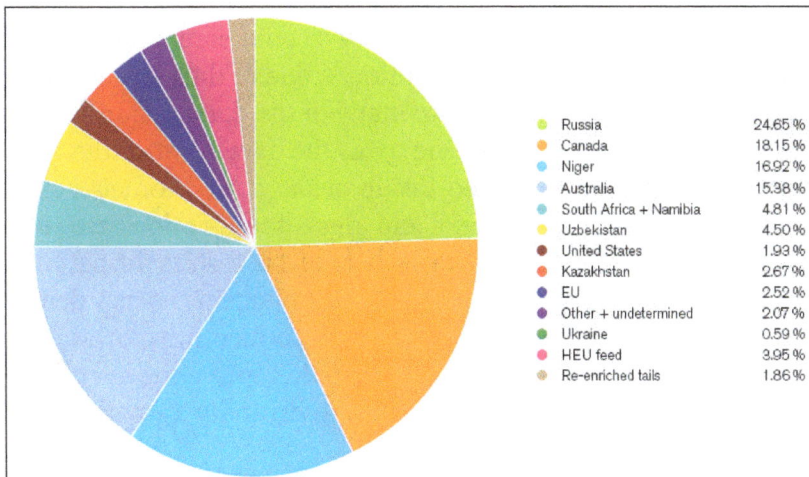

Russia	24.65 %
Canada	18.15 %
Niger	16.92 %
Australia	15.38 %
South Africa + Namibia	4.81 %
Uzbekistan	4.50 %
United States	1.93 %
Kazakhstan	2.67 %
EU	2.52 %
Other + undetermined	2.07 %
Ukraine	0.59 %
HEU feed	3.95 %
Re-enriched tails	1.86 %

Sources of uranium delivered to EU utilities in 2007.

Uranium for nuclear power is mined and enriched in diverse and "stable" countries. These include Canada (23% of the world's total in 2007), Australia (21%), Kazakhstan (16%) and more than 10 other countries. Uranium is mined and fuel is manufactured significantly in advance of need. Nuclear fuel is considered by some to be a relatively reliable power source, being more common in the Earth's crust than tin, mercury or silver, though a debate over the timing of peak uranium does exist.

Nuclear power reduces carbon emissions. Although a very viable resource, nuclear power can be a controversial solution because of the risks associated with it. Another factor in the debate with nuclear power is that many people or companies simply do not want any nuclear energy plant or radioactive waste near them.

Currently, nuclear power provides 13% of the world's total electricity. The most notable use of nuclear power within the United States is in U.S. Navy aircraft carriers and submarines, which have been exclusively nuclear-powered for several decades. These classes of ship provide the core of the Navy's power, and as such are the single most noteworthy application of nuclear power in that country.

Renewable Energy

The deployment of renewable technologies usually increases the diversity of electricity sources and, through local generation, contributes to the flexibility of the system and its resistance to central shocks. For those countries where growing dependence on imported gas is a significant energy security issue, renewable technologies can provide alternative sources of electric power as well as displacing electricity demand through direct heat production. Renewable biofuels for transport represent a key source of diversification from petroleum products.

As the resources that have been so crucial to survival in the world to this day start declining in numbers, countries will begin to realize that the need for renewable fuel sources will be as vital as ever. With the production of new types of energy, including solar, geothermal, hydro-electric, biofuel, and wind power. With the amount of solar energy that hits the world in one hour there is enough energy to power the world for one year. With the addition of solar panels all around the world a little less pressure is taken off the need to produce more oil.

Geothermal can potentially lead to other sources of fuel, if companies would take the heat from the inner core of the earth to heat up water sources we could essentially use the steam creating from the heated water to power machines, this option is one of the cleanest and efficient options. Hydro-electric which has been incorporated into many of the dams around the world, produces a lot of energy, and is very easy to produce the energy as the dams control the water that is allowed through seams which power turbines located inside of the dam. Biofuels have been researched using many different sources including ethanol and algae, these options are substantially cleaner than the consumption of petroleum. "Most life cycle analysis results for perennial and ligno-cellulosic crops conclude that biofuels can supplement anthropogenic energy demands and mitigate green house gas emissions to the atmosphere". Using oil to fuel transportation is a major source of green house gases, any one of these developments could replace the energy we derive from oil. Traditional fossil fuel exporters (e.g. Russia) struggle to diversify away from oil and develop renewable energy.

Energy Policy

Energy policy is the manner in which a given entity (often governmental) has decided to address issues of energy development including energy production, distribution and consumption. The attributes of energy policy may include legislation, international treaties, incentives to investment, guidelines for energy conservation, taxation and other public policy techniques. Energy is a core component of modern economies. A functioning economy requires not only labor and capital but also energy, for manufacturing processes, transportation, communication, agriculture, and more.

Concerning the term of energy policy, the importance of implementation of an eco-energy-oriented policy at a global level to address the issues of global warming and climate changes should be accentuated.

Although research is ongoing, the "human dimensions" of energy use are of increasing interest to business, utilities, and policymakers. Using the social sciences to gain insights into energy consumer behavior can empower policymakers to make better decisions about broad-based climate and energy options. This could facilitate more efficient energy use, renewable energy commercialization, and carbon emission reductions. Access to energy is also critical for basic social needs, such as lighting, heating, cooking, and health care. As a result, the price of energy has a direct effect on jobs, economic productivity and business competitiveness, and the cost of goods and services.

National Energy Policy

Measures used to Produce an Energy Policy

A national energy policy comprises a set of measures involving that country's laws, treaties and agency directives. The energy policy of a sovereign nation may include one or more of the following measures:

- Statement of national policy regarding energy planning, energy generation, transmission and usage.

- Legislation on commercial energy activities (trading, transport, storage, etc).

- Legislation affecting energy use, such as efficiency standards, emission standards.

- Instructions for state-owned energy sector assets and organizations.

- Active participation in, co-ordination of and incentives for mineral fuels exploration and other energy-related research and development policy command.

- Fiscal policies related to energy products and services (taxes, exemptions, subsidies).

- Energy security and international policy measures such as:

  ◦ International energy sector treaties and alliances,

  ◦ General international trade agreements,

  ◦ Special relations with energy-rich countries, including military presence and/or domination.

Frequently the dominant issue of energy policy is the risk of supply-demand mismatch. Current energy policies also address environmental issues, particularly challenging because of the need to reconcile global objectives and international rules with domestic needs and laws. Some governments state explicit energy policy, but, declared or not, each government practices some type of energy policy. Economic and energy modelling can be used by governmental or inter-governmental bodies as an advisory and analysis tool.

Factors within an Energy Policy

There are a number of elements that are naturally contained in a national energy policy, regardless of which of the above measures was used to arrive at the resultant policy. The chief elements intrinsic to an energy policy are:

- What is the extent of energy self-sufficiency for this nation?

- Where future energy sources will derive?

- How future energy will be consumed (e.g. among sectors)?

- What fraction of the population will be acceptable to endure energy poverty?

- What are the goals for future energy intensity, ratio of energy consumed to GDP?

- What is the reliability standard for distribution reliability?

- What environmental externalities are acceptable and are forecast?

- What form of "portable energy" is forecast (e.g. sources of fuel for motor vehicles)?

- How will energy efficient hardware (e.g. hybrid vehicles, household appliances) be encouraged?

- How can the national policy drive province, state and municipal functions?

- What specific mechanisms (e.g. taxes, incentives, manufacturing standards) are in place to implement the total policy?

- What future consequences there will be for national security and foreign policy ?

State, Province or Municipal Energy Policy

Even within a state it is proper to talk about energy policies in plural. Influential entities, such as municipal or regional governments and energy industries, will each exercise policy. Policy measures available to these entities are lesser in sovereignty, but may be equally important to national measures. In fact, there are certain activities vital to energy policy which realistically cannot be administered at the national level, such as monitoring energy conservation practices in the process of building construction, which is normally controlled by state-regional and municipal building codes (although can appear basic federal legislation).

Brazil

Brazil is the 10th largest energy consumer in the world and the largest in South America. At the same time, it is an important oil and gas producer in the region and the world's second largest ethanol fuel producer. The governmental agencies responsible for energy policy are the Ministry of Mines and Energy (MME), the National Council for Energy Policy (CNPE, in the Portuguese-language acronym), the National Agency of Petroleum, Natural Gas and Biofuels (ANP) and the National Agency of Electricity (ANEEL). State-owned companies Petrobras and Eletrobrás are the major players in Brazil's energy sector.

United States

Currently, the major issues in U.S. energy policy revolve around the rapidly growing production of domestic and other North American energy resources. The U.S. drive toward energy independence and less reliance on oil and coal is fraught with partisan conflict because these issues revolve around how best to balance both competing values, such as environmental protection and economic growth, and the demands of rival organized interests, such as those of the fossil fuel industry and of the newer renewable energy businesses.

European Union

Although the European Union has legislated, set targets, and negotiated internationally in the area of energy policy for many years, and evolved out of the European Coal and Steel Community, the concept of introducing a mandatory common European Union energy policy was only approved at the meeting of the European Council on October 27, 2005 in London. Following this the first policy proposals, *Energy for a Changing World*, were published by the European Commission, on January 10, 2007. The most well known energy policy objectives in the EU are 20/20/20 objectives, binding for all EU Member States. The EU is planning to increase the share of renewable energy in its final energy use to 20%, reduce greenhouse gases by 20% and increase energy efficiency by 20%.

Germany

In September 2010, the German government adopted a set of ambitious goals to transform their national energy system and to reduce national greenhouse gas emissions by 80 to 95% by 2050 (relative to 1990). This transformation become known as the *Energiewende*. Subsequently, the government decided to the phase-out the nation's fleet of nuclear reactors, to be complete by 2022. As of 2014, the country is making steady progress on this transition.

United Kingdom

The energy policy of the United Kingdom has achieved success in reducing energy intensity (but still really high), reducing energy poverty, and maintaining energy supply reliability to date. The United Kingdom has an ambitious goal to reduce carbon dioxide emissions for future years, but it is unclear whether the programs in place are sufficient to achieve this objective (the way to be so efficient as France is still hard). Regarding energy self sufficiency, the United Kingdom policy does not address this issue, other than to concede historic energy self sufficiency is currently ceasing to exist (due to the decline of the North Sea oil production). With regard to transport, the United Kingdom historically has a good policy record encouraging public transport links with cities, despite encountering problems with high speed trains, which have the potential to reduce dramatically domestic and short-haul European flights. The policy does not, however, significantly encourage hybrid vehicle use or ethanol fuel use, options which represent viable short term means to moderate rising transport fuel consumption. Regarding renewable energy, the United Kingdom has goals for wind and tidal energy. The White Paper on Energy, 2007, set the target that 20% of the UK's energy must come from renewable sources by 2020.

Soviet Union and Russia

The Soviet Union was the largest energy provider in the world until the late 1980s. Russia, one of the world's energy superpowers, is rich in natural energy resources, the world's leading net energy exporter, and a major supplier to the European Union. The main document defining the energy policy of Russia is the Energy Strategy, which initially set out policy for the period up to 2020, later was reviewed, amended and prolonged up to 2030. While Russia has also signed and ratified the Kyoto Protocol. Numerous scholars note that Russia uses its energy exports as a foreign policy instrument towards other countries.

India

The energy policy of India is characterized by trades between four major drivers:

- Rapidly growing economy, with a need for dependable and reliable supply of electricity, gas, and petroleum products;

- Increasing household incomes, with a need for affordable and adequate supply of electricity, and clean cooking fuels;

- Limited domestic reserves of fossil fuels, and the need to import a vast fraction of the gas, crude oil, and petroleum product requirements, and recently the need to import coal as well;

- Indoor, urban and regional environmental impacts, necessitating the need for the adoption of cleaner fuels and cleaner technologies.

In recent years, these challenges have led to a major set of continuing reforms, restructuring and a focus on energy conservation.

Thailand

The energy policy of Thailand is characterized by: 1) increasing energy consumption efficiency, 2) increasing domestic energy production, 3) increasing the private sector's role in the energy sector, 4) increasing the role of market mechanisms in setting energy prices. These policies have been consistent since the 1990s, despite various changes in governments. The pace and form of industry liberalization and privatization has been highly controversial.

Bangladesh

The first National Energy Policy (NEP) of Bangladesh was formulated in 1996 by the Ministry of Power, Energy and Mineral resources to ensure proper exploration, production, distribution and rational use of energy resources to meet the growing energy demands of different zones, consuming sectors and consumers groups on a sustainable basis. With rapid change of global as well as domestic situation, the policy was updated in 2004. The updated policy included additional objectives namely to ensure environmentally sound sustainable energy development programmes causing minimum damage to environment, to encourage public and private sector participation in the development and management of energy sector and to bring the entire country under electrification by the year 2020.

Australia

Australia's energy policy features a combination of coal power stations and hydro electricity plants. The Australian government has decided not to build nuclear power plants, although it is one of the world's largest producers of uranium.

Energy Crisis

An energy crisis is any significant bottleneck in the supply of energy resources to an economy. In literature, it often refers to one of the energy sources used at a certain time and place, in particular those that supply national electricity grids or those used as fuel in vehicles.

Industrial development and population growth have led to a surge in the global demand for energy in recent years. In the 2000s, this new demand — together with Middle East tension, the falling value of the U.S. dollar, dwindling oil reserves, concerns over peak oil, and oil price speculation — triggered the 2000s energy crisis, which saw the price of oil reach an all-time high of $147.30 a barrel in 2008.

Causes

Most energy crisis have been caused by localized shortages, wars and market manipulation. Some have argued that government actions like tax hikes, nationalisation of energy companies, and regulation of the energy sector, shift supply and demand of energy away from its economic equilibrium. However, the recent historical energy crisis listed below were not caused by such factors. Market failure is possible when monopoly manipulation of markets occurs. A crisis can develop due to industrial actions like union organized strikes and government embargoes. The cause may be over-consumption, aging infrastructure, choke point disruption or bottlenecks at oil refineries and port facilities that restrict fuel supply. An emergency may emerge during very cold winters due to increased consumption of energy.

The gasoline shortages of World War II brought about the resurgence of horse-and-wagon delivery.

Large fluctuations and manipulations in future derivatives can have a substantial impact on price. Large investment banks control 80% of oil derivatives as of May 2012, compared to 30% only a decade ago. This increase contributed to an improvement of global energy output from 117 687 TWh in 2000 to 143 851TWh in 2008. Limitations on free trade for derivatives could reverse this trend of growth in energy production. Kuwaiti Oil Minister Hani Hussein stated that "Under the supply and demand theory, oil prices today are not justified," in an interview with Upstream.

Pipeline failures and other accidents may cause minor interruptions to energy supplies. A crisis could possibly emerge after infrastructure damage from severe weather. Attacks by terrorists or militia on important infrastructure are a possible problem for energy consumers, with a successful strike on a Middle East facility potentially causing global shortages. Political events, for example, when governments change due to regime change, monarchy collapse, military occupation, and coup may disrupt oil and gas production and create shortages. Fuel shortage can also be due to the excess and useless use of the fuels.

Regular Gasoline Prices

Cents per Gallon

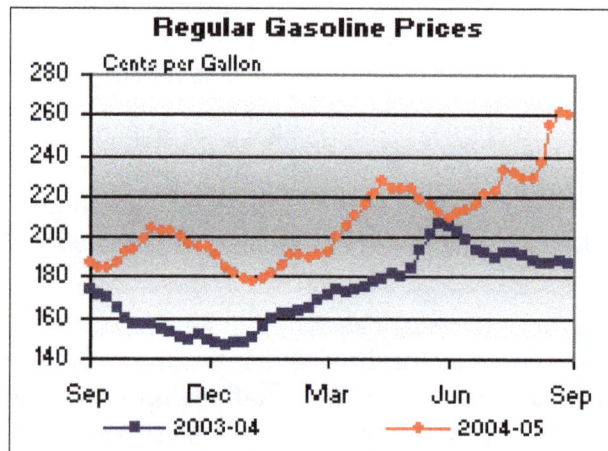

Various Historical Crises

- 1970s energy crisis - Caused by the peaking of oil production in major industrial nations (Germany, United States, Canada, etc.) and embargoes from other producers.

  - 1973 oil crisis - Caused by an OAPEC oil export embargo by many of the major Arab oil-producing states, in response to Western support of Israel during the Yom Kippur War.

  - 1979 oil crisis - Caused by the Iranian Revolution.

- 1990 oil price shock - Caused by the Gulf War.

- The 2000–2001 California electricity crisis - Caused by market manipulation by Enron and failed deregulation; resulted in multiple large-scale power outages.

- Fuel protests in the United Kingdom in 2000 were caused by a rise in the price of crude oil combined with already relatively high taxation on road fuel in the UK.

- North American natural gas crisis 2000-2008.

- 2004 Argentine energy crisis.

- North Korea has had energy shortages for many years.

- Zimbabwe has experienced a shortage of energy supplies for many years due to financial mismanagement.

- Political riots occurring during the 2007 Burmese anti-government protests were sparked by rising energy prices.

Kuwait's Al Burqan Oil Field, the world's second largest oil field.

- 2000s energy crisis - Since 2003, a rise in prices caused by continued global increases in petroleum demand coupled with production stagnation, the falling value of the U.S. dollar, and a myriad of other secondary causes.

- 2008 Central Asia energy crisis, caused by abnormally cold temperatures and low water levels in an area dependent on hydroelectric power. At the same time the South African President was appeasing fears of a prolonged electricity crisis in South Africa.

- In February 2008 the President of Pakistan announced plans to tackle energy shortages that were reaching crisis stage, despite having significant hydrocarbon reserves,. In April 2010, the Pakistani government announced the Pakistan national energy policy, which extended the official weekend and banned neon lights in response to a growing electricity shortage.

- South African electrical crisis. The South African crisis led to large price rises for platinum in February 2008 and reduced gold production.

- China experienced severe energy shortages towards the end of 2005 and again in early 2008. During the latter crisis they suffered severe damage to power networks along with diesel and coal shortages. Supplies of electricity in Guangdong province, the manufacturing hub of China, are predicted to fall short by an estimated 10 GW. In 2011 China was forecast to have a second quarter electrical power deficit of 44.85 - 49.85 GW.

- Nepal experienced severe energy crisis in 2015 when India created an economic blockade to Nepal. Nepal faced the shortages of various kinds of petroleum products and food materials which affected severely on Nepal's economy.

- The Gaza electricity crisis is a result of the tensions between Hamas, who rules the Gaza Strip, and the Palestinian Authority/Fatah, who rules the West Bank over custom tax revenue, funding of the Gaza Strip, and political authority. Residents receive electricity for a few hours a day on a rolling blackout schedule.

Emerging Oil Shortage

"Peak oil" is the period when the maximum rate of global petroleum extraction is reached, after which the rate of production enters terminal decline. It relates to a long-term decline in the available supply of petroleum. This, combined with increasing demand, significantly increases the worldwide prices of petroleum derived products. Most significant is the availability and price of liquid fuel for transportation.

The US Department of Energy in the Hirsch report indicates that "The problems associated with world oil production peaking will not be temporary, and past 'energy crisis' experience will provide relatively little guidance."

Mitigation Efforts

To avoid the serious social and economic implications a global decline in oil production could entail, the 2005 Hirsch report emphasized the need to find alternatives, at least ten to twenty years before the peak, and to phase out the use of petroleum over that time. Such mitigation could include energy conservation, fuel substitution, and the use of unconventional oil. Because mitigation can reduce the use of traditional petroleum sources, it can also affect the timing of peak oil and the shape of the Hubbert curve.

Energy policy may be reformed leading to greater energy intensity, for example in Iran with the 2007 Gas Rationing Plan in Iran, Canada and the National Energy Program and in the USA with the *Energy Independence and Security Act of 2007* also called the *Clean Energy Act of 2007*. Another mitigation measure is the setup of a cache of secure fuel reserves like the United States Strategic Petroleum Reserve, in case of national emergency. Chinese energy policy includes specific targets within their 5-year plans.

Andrew McKillop has been a proponent of a contract and converge model or capping scheme, to mitigate both emissions of greenhouse gases and a peak oil crisis. The imposition of a carbon tax would have mitigating effects on an oil crisis. The Oil Depletion Protocol has been developed by Richard Heinberg to implement a powerdown during a peak oil crisis. While many sustainable development and energy policy organisations have advocated reforms to energy development from the 1970s, some cater to a specific crisis in energy supply including Energy-Quest and the International Association for Energy Economics. The Oil Depletion Analysis Centre and the Association for the Study of Peak Oil and Gas examine the timing and likely effects of peak oil.

Ecologist William Rees believes that:

> " To avoid a serious energy crisis in coming decades, citizens in the industrial countries should actually be urging their governments to come to international agreement on a persistent, orderly, predictable, and steepening series of oil and natural gas price hikes over the next two decades."

Due to a lack of political viability on the issue, government mandated fuel prices hikes are unlikely and the unresolved dilemma of fossil fuel dependence is becoming a wicked problem. A global soft energy path seems improbable, due to the rebound effect. Conclusions that the world is heading towards an unprecedented large and potentially devastating global energy crisis due to a decline in the availability of cheap oil lead to calls for a decreasing dependency on fossil fuel.

Other ideas concentrate on design and development of improved, energy-efficient urban infrastructure in developing nations. Government funding for alternative energy is more likely to increase during an energy crisis, so too are incentives for oil exploration. For example, funding for research into inertial confinement fusion technology increased during the 1970s.

Kirk Sorensen and others have suggested that additional nuclear power plants, particularly liquid fluoride thorium reactors have the energy density to mitigate global warming and replace the energy from peak oil, peak coal and peak gas. The reactors produce electricity and heat so much of the transportation infrastructure should move over to electric vehicles. However, the high process heat of the molten salt reactors could be used to make liquid fuels from any carbon source.

2010s Oil Glut

Rather counterintuitively, the world economy has had to deal with the unforeseen consequences of the 2015-2016 oil glut also known as 2010s oil glut, a major energy crisis that took many experts by surprise. This oversupply crisis started with a considerable time-lag, more than six years after the beginning of the Great Recession: "the price of oil [had] stabilized at a relatively high level (around $100 a barrel) unlike all previous recessionary cycles since 1980 (start of First Persian Gulf War). But nothing guarantee[d] such price levels in perpetuity".

Social and Economic Effects

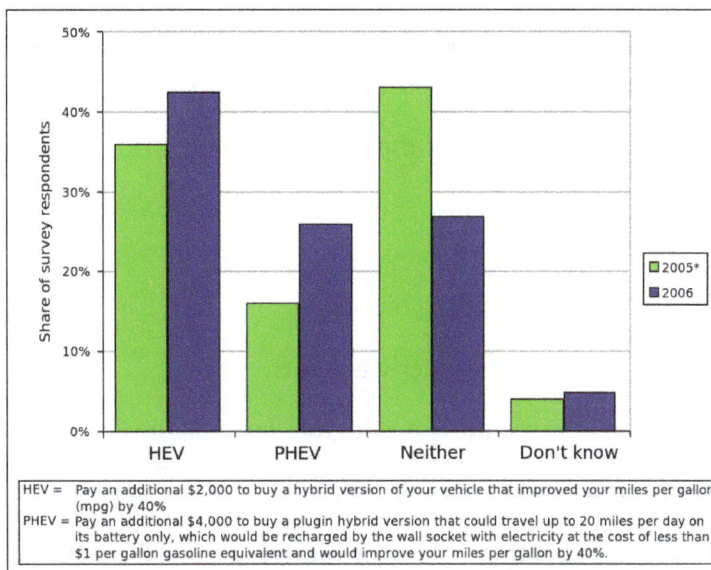

In 2006, survey respondents in the United States were willing to pay more for a plug-in hybrid car.

The macroeconomic implications of a supply shock-induced energy crisis are large, because energy is the resource used to exploit all other resources. Oil price shocks can affect the rest of the economy through delayed business investment, sectoral shifts in the labor market, or monetary policy responses. When energy markets fail, an energy shortage develops. Electricity consumers may experience intentionally engineered rolling blackouts during periods of insufficient supply or unexpected power outages, regardless of the cause.

Industrialized nations are dependent on oil, and efforts to restrict the supply of oil would have an adverse effect on the economies of oil producers. For the consumer, the price of natural gas, gasoline (petrol) and diesel for cars and other vehicles rises. An early response from stakeholders is the call for reports, investigations and commissions into the price of fuels. There are also movements towards the development of more sustainable urban infrastructure.

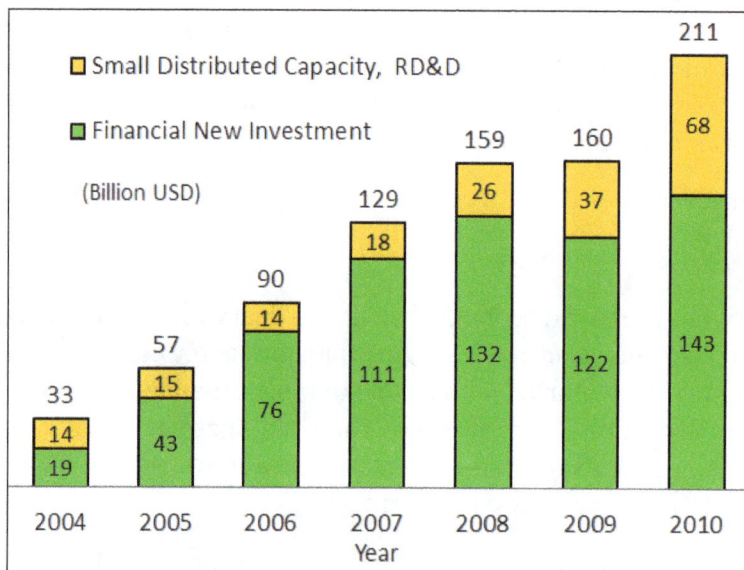

Global New Investments in Renewable Energy.

In the market, new technology and energy efficiency measures become desirable for consumers seeking to decrease transport costs. Examples include:

- In 1980 Briggs & Stratton developed the first gasoline hybrid electric automobile; also are appearing plug-in hybrids.

- The growth of advanced biofuels.

- Innovations like the Dahon, a folding bicycle.

- Modernized and electrifying passenger transport.

- Railway electrification systems and new engines such as the Ganz-Mavag locomotive.

- Variable compression ratio for vehicles.

Other responses include the development of unconventional oil sources such as synthetic fuel from places like the Athabasca Oil Sands, more renewable energy commercialization and use of alternative propulsion. There may be a Relocation trend towards local foods and possibly microgeneration, solar thermal collectors and other green energy sources.

Tourism trends and gas-guzzler ownership varies with fuel costs. Energy shortages can influence public opinion on subjects from nuclear power plants to electric blankets. Building construction techniques—improved insulation, reflective roofs, thermally efficient windows, etc.—change to reduce heating costs.

Crisis Management

An electricity shortage is felt most acutely in heating, cooking, and water supply. Therefore, a sustained energy crisis may become a humanitarian crisis.

If an energy shortage is prolonged a crisis management phase is enforced by authorities. Energy audits may be conducted to monitor usage. Various curfews with the intention of increasing energy conservation may be initiated to reduce consumption. For example, to conserve power during the Central Asia energy crisis, authorities in Tajikistan ordered bars and cafes to operate by candlelight.

In the worst kind of energy crisis energy rationing and fuel rationing may be incurred. Panic buying may beset outlets as awareness of shortages spread. Facilities close down to save on heating oil; and factories cut production and lay off workers. The risk of stagflation increases.

Energy Efficiency

Energy efficiency simply means using less energy to perform the same task – that is, eliminating energy waste. Energy efficiency brings a variety of benefits: reducing greenhouse gas emissions, reducing demand for energy imports, and lowering our costs on a household and economy-wide level. While renewable energy technologies also help accomplish these objectives, improving energy efficiency is the cheapest – and often the most immediate – way to reduce the use of fossil fuels. There are enormous opportunities for efficiency improvements in every sector of the economy, whether it is buildings, transportation, industry, or energy generation.

Buildings

Building designers are looking to optimize building efficiency and then incorporate renewable energy technologies, leading to the creation of zero-energy buildings. Changes in existing buildings can also be made to reduce energy usage and costs. These may include small steps, such as choosing LED light bulbs and energy efficient appliances, or larger efforts such as upgrading insulation and weatherization.

Energy Generation and Distribution

Combined heat and power systems capture the "waste" heat from power plants and use it to provide heating, cooling, and/or hot water to nearby buildings and facilities. This increases the energy efficiency of power generation from approximately 33 percent to up to 80 percent. The smart grid is another system that will improve the efficiency of electric generation, distribution, and consumption.

Community Design

Neighborhoods that are designed with mixed use developments and safe, accessible options for walking, biking, and public transportation are key to reducing the need for personal vehicle travel.

Vehicles

More energy efficient vehicles require less fuel to cover a given distance. This generates fewer emissions, and makes them significantly less expensive to operate. Plug-in hybrids and fully electric vehicles are particularly fuel efficient.

Freight

Freight can be moved more efficiently by improving the efficiency of rail and truck transportation and by shifting long-distance freight transport from trucks to rail.

Human Behavior

The four strategies above improve energy efficiency primarily through technology and design. However, the way people use these technologies will significantly impact their effectiveness. What impact can a highly efficient technology have if households and businesses are not motivated to buy, install, and/or activate it? How does driving behavior and unnecessary idling impact gas mileage? How many people will use public transportation if there is a cultural stigma against it? Research has shown that 30 percent of the potential energy savings of high efficiency technologies is lost due to a variety of social, cultural, and economic factors. Addressing these factors is also an important component of making our economy more energy efficient.

Need for Energy efficiency

Energy consumption has grown incredibly fast over the last few decades. We are in danger of using up the planet's natural resources, of destroying vital habitats and polluting the air we need to breathe.

Energy efficiency is a way of managing and limiting this growth in energy consumption, to save wildlife habitats, safeguard the planet, and make sure there is energy left for future generations.

Importance of Energy Efficiency

Energy efficiency is playing an increasingly vital role in our lives, for three main reasons.

Environment

The more energy we use, the more carbon emissions are pumped into the atmosphere and the more our reserves of natural resources such as oil, coal and gas are depleted. We need to reduce our reliance on these energy sources, and one way to do that is to make sure we all use energy as efficiently as possible.

Global Economy

The global economy is based heavily on oil and gas, and as these resources dwindle their cost will increase, causing financial imbalances around the world and resulting in energy poverty in many areas of society.

Your bank balance

Nobody wants to pay more than they have to for everyday necessities like heating and hot water, so it makes sense to be energy efficient. That way you fulfil your energy needs while paying as little as possible.

Ways to be more Energy Efficient

If you want to know how to be more energy efficient, the first place to start is to make sure your home and all your electrical appliances work as efficiently as possible. It isn't energy efficient to throw out all your old products and replace them with new, energy-saving items; instead, wait until they wear out and then replace them.

Use Less Electricity

The first rule of saving electricity is: don't leave appliances on standby. Of course you need to leave your fridge and freezer on full time, and maybe your alarm system – and you may need to leave the TV or satellite box on to record your favourite programmes. But for practically everything else electrical: when you're not using them, switch them off at the wall.

Rechargeable batteries can also be a good choice. Just make sure you get top-quality ones that will last and hold their charge. You could also invest in a 'smart charger' that can prevent your batteries from overcharging.

If you want to make your laptop more energy efficient, you can get an Ecobutton. It plugs into a USB port, and flashes to remind you to press if you decide to stop using your computer for a while. It can then put your computer into its most efficient energy-saving mode. When you log on again, you can see on screen how much money and CO_2 you've saved.

Switch to Energy-saving Light Bulbs

Energy-efficient bulbs use up to 80% less electricity than traditional light bulbs and can keep going for ten times as long.

Just one energy-saving light bulb could save you approximately £2.50 per year – and this could rise to as much as £6 for brighter bulbs or any you leave on for several hours each day. So if you replace each of the standard bulbs in your home (when they stop working) with energy-saving bulbs, you could reduce your annual energy bill by as much as £37 and cut 135 kg of CO_2 off your carbon footprint. That's about the equivalent of taking a train from London to Glasgow and back, or using nearly 43 litres of petrol.

Cut Down the Cost of Heating your Home

Hot water and heating make up around four-fifths of most fuel bills in the UK, so increasing the energy efficiency of your heating system can make a big difference to your utility bills.

If you've had your boiler for more than 15 years, it's probably time to consider upgrading to a newer, more eco-friendly model. You could reduce your heating bills by up to a quarter if you replace a creaky old G-rated boiler with a new A-rated condensing boiler – as long as you use it wisely and control it effectively. Make sure your thermostat and boiler are communicating properly, fit individual thermostats on radiators, and get a control system that lets you switch off the heating remotely if necessary.

Insulate your roof and walls, install double glazing, stop draughts, update your heating system and take a bit more care about keeping doors and windows shut. You could also cut up to 10% off your heating bills if you lower your heating thermostat by 1°C and put on an extra jumper or fleece instead. There's no need to have the thermostat on a hot water tank any higher than 60 °C/140 °F.

Energy Monitoring and Targeting

Energy monitoring and targeting (M&T) is an energy efficiency technique based on the standard management axiom stating that "you cannot manage what you cannot measure". M&T techniques provide energy managers with feedback on operating practices, results of energy management projects, and guidance on the level of energy use that is expected in a certain period. Importantly, they also give early warning of unexpected excess consumption caused by equipment malfunctions, operator error, unwanted user behaviours, lack of effective maintenance and the like.

The foundation of M&T lies in determining the normal relationships of energy consumptions to relevant driving factors (HVAC equipment, production though puts, weather, occupancy available daylight, etc.) and the goal is to help business managers:

- Identify and explain excessive energy use.

- Detect instances when consumption is unexpectedly higher or lower than would usually have been the case.

- Visualize energy consumption trends (daily, weekly, seasonal, operational).

- Determine future energy use and costs when planning changes in the business.

- Diagnose specific areas of wasted energy.

- Observe how changes to relevant driving factors impact energy efficiency.

- Develop performance targets for energy management programs.

- Manage energy consumption, rather than accept it as a fixed cost.

The ultimate goal is to reduce energy costs through improved energy efficiency and energy management control. Other benefits generally include increased resource efficiency, improved production budgeting and reduction of greenhouse gas (GHG) emissions.

Goals and Benefits

Throughout the numerous M&T projects implemented since the 1980s, a certain number of benefits have proved to be recurrent:

- Energy cost savings: Generally 5% of the original energy expenses. Carbon Trust has conducted a study over 1000 small businesses and has concluded that on average an organization could save 5%.

- Reduction in GHG emissions: Lower energy consumption helps reduce emissions.

- Financing: Measured energy reductions help obtain grants for energy efficiency projects.

- Improved product and service costing: Sub-metering allows the division of the energy bill between the different processes of an industry, and can be calculated as a production cost.

- Improved budgeting: M&T techniques can help forecast energy expenses in the case of changes in the business, for example.

- Waste avoidance: Helps diagnose energy waste in any process.

Technique

Key Principles of Energy Monitoring and Targeting

Monitoring and Targeting techniques rely on three main principles, which form a constant feedback cycle, therefore improving control of energy use.

Monitoring

Monitoring information of energy use, in order to establish a basis for energy management and explain deviations from an established pattern. Its primary goal is to maintain said pattern, by providing all the necessary data on energy consumption, as well as certain driving factors, as identified during preliminary investigation (production, weather, etc).

Reporting

The final principle is the one which enables ongoing control of energy use, achievement of targets and verification of savings: reports must be issued to the appropriate managers. This in turn allows decision-making and actions to be taken in order to achieve the targets, as well as confirmation or denial that the targets have been reached.

Procedures

Before the M&T measures themselves are implemented, a few preparatory steps are necessary. First of all, key energy consumers on the site must be identified. Generally, most of the energy consumption is concentrated in a small number of processes, like heating, or certain machinery. This normally requires a certain survey of the building and the equipment to estimate their energy consumption level.

It is also necessary to assess what other measurements will be required to analyze the consumption appropriately. This data will be used to chart against the energy consumption: these are underlying factors which influence the consumption, often production (for industry processes) or exterior temperature (for heating processes), but may include many other variables.

Once all variables to be measured have been established, and the necessary meters installed, it is possible to initiate the M&T procedures.

Measure

The first step is to compile the data from the different meters. Low-cost energy feedback displays have become available. The frequency at which the data is compiled varies according to the desired reporting interval, but can go once every 30 seconds to once every 15 minutes. Some measurements can be taken directly from the meters, others must be calculated. These different measurements are often called streams or channels.

Driving factors such as production or degree days also constitute streams and must be collected at intervals to match.

Define the Base-line

The data compiled must then be plotted on a graph in order to define the general consumption base-line. Consumption rates are plotted in a scatter plot against production or any other variable previously identified, and the best fit line is identified. This graph is the image of the business' average energy performance, and conveys a lot of information:

- The y-intercept gives the minimal consumption in the absence of the variable (no production, zero degree-day). This is the base load of the system, the minimal consumption when it is not operating.

- The slope represents the relationship between the consumption and the previously identified variable. This represents the efficiency of the process.

- The scatter is the degree of variability of the consumption with operational factors.

The slope is not used quite as often for M&T purposes. However, a high y-intercept can mean that there is a fault in the process, causing it to use too much energy with no performance, unless there are specific distinctive features which lead to high base loads. Very scattered points, on the other hand, may reflect other significant factors playing in the variation of the energy consumption, other than the one plotted in the first place, but it can also be the illustration of a lack of control over the process.

Monitor Variations

The next step is to monitor the difference between the expected consumption and the actual measured consumption. One of the tools most commonly used for this is the CUSUM, which is the CUmulative SUM of differences. This consists in first calculating the difference between the expected and actual performances (the best fit line previously identified and the points themselves).

The CUSUM can then be plotted against time on a new graph, which then yields more information for the energy efficiency specialist. Variances scattered around zero usually mean that the process is operating normally. Marked variations, increasing or decreasing steadily usually reflect a modification in the conditions of the process.

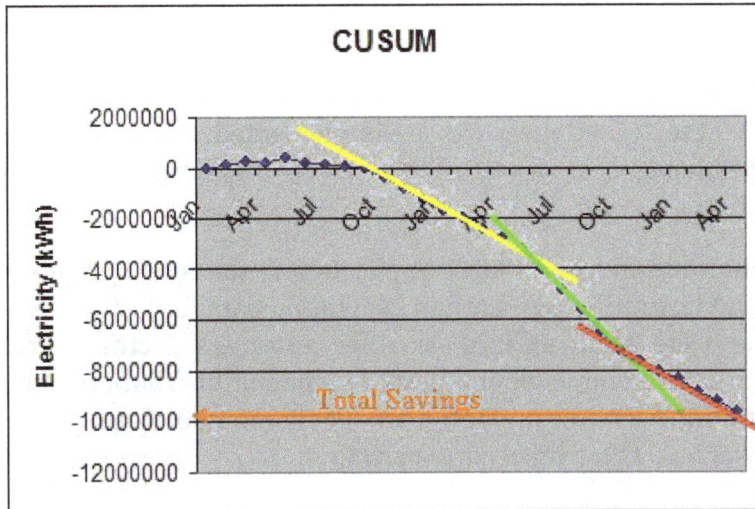

In the case of the CUSUM graph, the slope becomes very important, as it is the main indicator of the savings achieved. A slope going steadily down indicates steady savings. Any variation in the slope indicates a change in the process. For example, in the graph on the right, the first section indicated no savings. However, in September (beginning of the yellow line), an energy efficiency measure must have been implemented, as savings start to occur. The green line indicates an increase in the savings (as the slope is becoming steeper), whereas the red line must reflect a modification in the process having occurred in November, as savings have decreased slightly.

Identify Causes

Energy efficiency specialists, in collaboration with building managers, will decipher the CUSUM graph and identify the causes leading to variations in the consumption. This can be a change in behaviour, a modification to the process, different exterior conditions, etc. These changes must be monitored and the causes identified in order to promote and enhance good behaviour, and discourage bad ones.

Set Targets

Once the base line has been established, and causes for variations in energy consumption have been identified, it is time to set targets for the future. Now with all this information in hand, the targets are more realistic, as they are based on the building's actual consumption. Targeting consists in two main parts: the measure to which the consumption can be reduced, and the timeframe during which the compression will be achieved.

A good initial target is the best fit line identified during step 2. This line represents the average historical performance. Therefore, keeping all consumption below or equal to the historical average is an achievable target, yet remains a challenge as it involves eliminating high consumption peaks.

Some companies, as they improve their energy consumption, might even decide to bring their average performance down to their historical best. This is considered a much more challenging target.

Monitor Results

This brings us back to step 1: measure consumption. One of the specificities of M&T is that it is an ongoing process, requiring constant feedback in order to consistently improve performance. Once the targets are set and the desired measures are implemented, repeating the procedure from the start ensures that the managers are aware of the success or failure of the measures, and can then decide on further action.

An example with some features of an M&T application is the ASU Campus Metabolism, which provides real-time and historic energy use and generation data for facilities of Arizona State University on a public web site. Many utilities also offer customers electric interval data monitoring services. Xcel Energy is an example of an investor owned utility that offers its customer electric and natural gas monitoring services under the product name InfoWise from Xcel Energy which is administered by Power TakeOff, a third party partner.

Energy Efficiency in Transport

Safe, equitable, and energy-efficient urban transport can help achieve multiple health and sustainability goals. Shifting urban design and infrastructure investments into public transport networks that prioritize rapid bus transit or light rail over private vehicles can reduce the long-term trajectory of both air pollution and climate emissions generated by private transport – and improve health equity by providing those lacking cars with better mobility.

Diesel vehicles are the heaviest source of particulate (PM) emissions, including black carbon climate pollutants (a short-lived climate pollutant (SLCP) that is a component of particulate pollution).

Low-sulphur diesel fuels and low-emissions vehicles, as well as a modal shift to public transport and non-motorized modes, both are essential in order to immediately reduce pollution and SLCP climate emissions. In developing cities, in particular, the absence of strong urban rapid transit and non- motorized transit systems, means that improvements in vehicle technologies are typically overtaken by increasing vehicle traffic – driving up pollution to previous levels and perpetuating a trajectory of higher short-lived (black carbon) and long-lived (CO_2) climate emissions.

This is a common cycle in many rapidly-growing low and middle income cities of Africa and Asia today, which face strong pressures for more travel – and weak public transport systems.

Complementary walking and cycling infrastructures are comparatively easy and inexpensive for local authorities to develop – when the political will exists. These can immediately reduce injury risks for a very large proportion of urban dwellers. For instance, in Nairobi it is estimated that some 40% of daily trips are on foot and another 40% of travel is via informal and poorly organized "matatus," or shared taxi systems – while only about 9% of travel is by private vehicles. As in most of Africa, no formal air quality monitoring system exists in Nairobi, however, research studies have attributed much of the city's air pollution to traffic, with reported PM2.5 air pollution levels several times over WHO guideline limits.

This illustrates how, over time, investments in rapid transit and non-motorized travel systems can help support healthy physical activity and further reduce air pollution and climate emissions with zero- emissions transport modes, as urban populations become more mobile.

Compact cities served by transit and dedicated walking and cycling networks are more energy- efficient and safer for pedestrians and cyclists. Long-term studies in cities as diverse as Shanghai and Copenhagen, studies have found a 30% lower annual mortality risk among cycle commuters – the gains in life expectancy from improved physical activity in these cities also outweighed increased exposures to injury and pollution. Cities built around transit and active transport also offer efficient and equitable access to jobs, health facilities, and other urban services; such transportation infrastructure is particularly important to youth, elderly, disabled, and low-income groups.

Traditional vehicle-based strategies foster sprawl due to the needs for larger roads and expanses of parking between buildings. As cities expand horizontally, to accommodate road and parking infrastructure needs, public transport becomes increasingly inefficient as does non-motorized transport, due to longer urban trips. New roads induce more vehicle travel, and progressively longer urban trips, in a vicious cycle. Sustainable transport solutions are therefore crucial for urban planning and design in order to increase accessibility without increasing travel times, pollution, and environmental risks.

References

- Energyconsumption: conserve-energy-future.com, Retrieved 30 April, 2019

- "Historical Statistics of Japan". Japan Ministry of Internal Affairs and Communications. Retrieved 3 April 2007

- Vincent, Jeffrey (February 2000). "Green accounting: from theory to practice". Environment and Development Economics. 5: 13–24. Doi:10.1017/S1355770X00000024

- "Consumption by fuel, 1965–2008". Statistical Review of World Energy 2009. BP. 8 June 2009. Archived from the original (XLS) on 26 July 2013. Retrieved 24 October 2009

- "IEA estimates $48tn investments till 2035 to meet global energy demands". Bloomberg News. Retrieved 4 June 2014

- What-is-energy-efficiency, energy-guides, guides: ovoenergy.com, Retrieved 29 March, 2019

- West, J (2011). "Decreasing metal ore grades: are they really being driven by the depletion of high-grade deposits?". J Ind Ecol. 15 (2): 165–168. Doi:10.1111/j.1530-9290.2011.00334.x

- F. William Engdahl (Mar 18, 2012). "Behind Oil Price Rise: Peak Oil or Wall Street Speculation?". Axis of Logic. Retrieved 21 March 2012

- Description, energy-efficiency: eesi.org, Retrieved 16 January, 2019

Strategies for Energy Conservation

4

- **Energy Conservation Techniques**
- **Energy Development**
- **Energy Recovery**
- **Energy Audit**
- **Energy Harvesting**
- **Alternative Energy**
- **Green Bulding**
- **Green Computing**
- **Passive House**
- **Zero Energy Building**
- **Smart Grid**

The major strategies of energy conservation include energy harvesting, energy recovery, green building, zero energy building, energy development, green computing, and using smart grids, alternate energy and passive houses. These strategies of energy conservation have been thoroughly discussed in this chapter.

Energy Conservation Techniques

Energy conservation is not about making limited resources last as long as they can, that would mean that you are doing nothing more than prolong a crisis until you finally run out of energy resources all together. Conservation is the process of reducing demand on a limited supply and enabling that supply to begin to rebuild itself. Many times the best way of doing this is to replace the energy used with an alternate.

In the case of fossil fuels, the conservation also can include finding new ways to tap into the Earth's supply so that the commonly used oil fields are not drained completely. This allows for those fields to replenish themselves more. This is not a process that happens overnight, when you are talking about replenishing natural resources you are talking about alleviating excess demand on the supply in 100's of years' time to allow nature to recover.

Need for Energy Conservation

Without energy conservation, the world will deplete its natural resources. While some people don't see that as an issue because it will take many decades to happen and they foresee that by the time the natural resource is gone there will be an alternative; the depletion also comes at the cost of creating an enormous destructive waste product that then impacts the rest of life. The goal with energy conservation techniques is reduce demand, protect and replenish supplies, develop and use alternative energy sources, and to clean up the damage from the prior energy processes.

Practical Methods of Energy Conservation

Below are energy conservation techniques that can help you to reduce your overall carbon footprint and save money in the long run.

1. Install CFL Lights: Try replacing incandescent bulbs in your home with CFL bulbs. CFL bulbs cost more upfront but last 12 times longer than regular incandescent bulbs. CFL bulbs will not only save energy but over time you end up saving money.

2. Lower the Room Temperature: Even a slight decrease in room temperature lets say by only a degree or two, can result in big energy savings. The more the difference between indoor and outdoor temperature, the more energy it consumes to maintain room temperature. A more smarter and comfortable way of doing this is to buy a programmable thermostat.

3. Fix Air Leaks: Proper insulation will fix air leaks that could be costing you. During winter months, you could be letting out a lot of heat if you do not have a proper insulation. You can fix those leaks yourself or call an energy expert to do it for you.

4. Use Maximum Daylight: Turn off lights during the day and use daylight as much as possible.

This will reduce the burden on the local power grid and save you good amount of money in the long run.

5. Get Energy Audit Done: Getting energy audit done by hiring an energy audit expert for your home is an energy conservation technique that can help you conserve energy and save good amount of money every month. Home energy audit is nothing but a process that helps you to identify areas in your home where it is losing energy and what steps you can take to overcome them. Implement the tips and suggestions given by those energy experts and you might see some drop in your monthly electricity bill.

6. Use Energy Efficient Appliances: When planning to buy some electrical appliances, prefer to buy one with Energy Star rating. Energy efficient appliances with Energy Star rating consume less energy and save you money. They might cost you more in the beginning but it is much more of an investment for you.

7. Drive Less, Walk More and Carpooling: Yet another energy conservation technique is to drive less and walk more. This will not only reduce your carbon footprint but will also keep you healthy as walking is a good exercise. If you go to office by car and many of your colleagues stay nearby, try doing carpooling with them. This will not only bring down your monthly bill you spend on fuel but will also make you socially more active.

8. Switch Off Appliances when Not in Use: Electrical appliances like coffee machine, idle printer, desktop computer keep on using electricity even when not in use. Just switch them off if you don't need them immediately.

9. Plant Shady Landscaping: Shady landscaping outside your home will protect it from intense heat during hot and sunny days and chilly winds during the winter season. This will keep your home cool during summer season and will eventually turn to big savings when you calculate the amount of energy saved at the end of the year.

10. Install Energy Efficient Windows: Some of the older windows installed at our homes aren't energy efficient. Double panel windows and other vinyl frames are much better than single pane windows. Choosing correct blinds can save on your power bills.

Other Energy Conservation Techniques

The other few energy conservation techniques may surprise you. While there are practical methods such as insulation, changing light sources, using alternate fuels and carpooling rather than walking – understand the 6 core techniques beneath them will show you more about what to do in life.

1. Education: Education is probably the most powerful of the energy conservation techniques that can be used. Education is about more than teaching people the importance of conservation, it is about showing the alternative choices that can be used in construction, manufacturing and other processes.

2. Zero Energy Balance: Zero Energy Balance is more than techniques of conserving energy in green construction. It is a process of re-evaluating and retrofitting manufacturing and commercial operations so that they can harvest and store energy, as well as take and replace it onto the grid to relieve brown out stresses.

3. Alternative Power: There are more processes that are starting to use alternative power and fuel sources in many different areas of life. The use of alternative power is one of the most key energy conservation techniques because almost all of the transition models require that the existing processes be upgraded or replaced to more energy efficient models too.

4. Cap and Trade Agreements: Cap and trade agreements are used as part of the process of regulating and conserving consumption and pollution for manufacturing industries. The companies are "allowed" a certain emission rate which they can bid buy to extend. The extension bid is then used for compensating projects. While this may not seem like it is directly related to energy conservation it is very much at its core.

5. Reduced Demand: There are numerous initiatives that are working to reduce the overall demand on the energy resources of the world. This can range everywhere from education programs to changing the type of required insulation in new construction.

6. Research and Development: Continued funding of research and development projects in the energy conservation field is how we discover the changes that can be made to reduce consumption and discover renewable methods to provide us with the energy that modern life requires. It should be one of the energy conservation techniques that are most valued as it is what holds the promise for leading to a solution to the world's energy crisis.

Energy Development

Energy development is the field of activities focused on obtaining sources of energy from natural resources. These activities include production of conventional, alternative and renewable sources of energy, and for the recovery and reuse of energy that would otherwise be wasted. Energy conservation and efficiency measures reduce the demand for energy development, and can have benefits to society with improvements to environmental issues.

Societies use energy for transportation, manufacturing, illumination, heating and air conditioning, and communication, for industrial, commercial, and domestic purposes. Energy resources may be classified as primary resources, where the resource can be used in substantially its original form, or as secondary resources, where the energy source must be converted into a more conveniently usable form. Non-renewable resources are significantly depleted by human use, whereas renewable resources are produced by ongoing processes that can sustain indefinite human exploitation.

Thousands of people are employed in the energy industry. The conventional industry comprises the petroleum industry, the natural gas industry, the electrical power industry, and the nuclear industry. New energy industries include the renewable energy industry, comprising alternative and sustainable manufacture, distribution, and sale of alternative fuels.

Classification of Resources

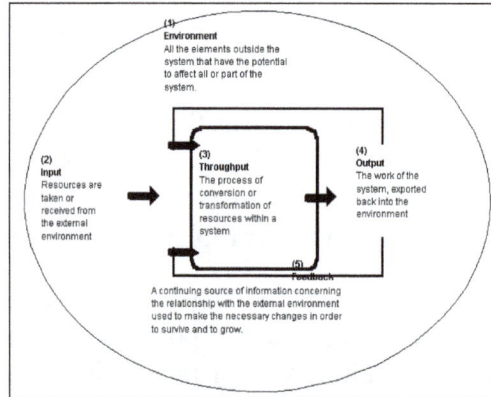

Open System Model (basics).

Energy resources may be classified as primary resources, suitable for end use without conversion to another form, or secondary resources, where the usable form of energy required substantial conversion from a primary source. Examples of primary energy resources are wind power, solar power, wood fuel, fossil fuels such as coal, oil and natural gas, and uranium. Secondary resources are those such as electricity, hydrogen, or other synthetic fuels.

Another important classification is based on the time required to regenerate an energy resource. "Renewable" resources are those that recover their capacity in a time significant by human needs. Examples are hydroelectric power or wind power, when the natural phenomena that are the primary source of energy are ongoing and not depleted by human demands. Non-renewable resources are those that are significantly depleted by human usage and that will not recover their potential significantly during human lifetimes. An example of a non-renewable energy source is coal, which does not form naturally at a rate that would support human use.

Fossil Fuels

The Moss Landing Power Plant in California is a fossil-fuel power station that burns natural gas in a turbine to produce electricity.

Fossil fuel (*primary non-renewable fossil*) sources burn coal or hydrocarbon fuels, which are the remains of the decomposition of plants and animals. There are three main types of fossil fuels: coal, petroleum, and natural gas. Another fossil fuel, liquefied petroleum gas (LPG), is principally derived from the production of natural gas. Heat from burning fossil fuel is used either directly for space heating and process heating, or converted to mechanical energy for vehicles, industrial processes, or electrical power generation. These fossil fuels are part of the carbon cycle and thus allow stored solar energy to be used today.

The use of fossil fuels in the 18th and 19th Century set the stage for the Industrial Revolution.

Fossil fuels make up the bulk of the world's current primary energy sources. In 2005, 81% of the world's energy needs was met from fossil sources. The technology and infrastructure already exist for the use of fossil fuels. Liquid fuels derived from petroleum deliver a great deal of usable energy per unit of weight or volume, which is advantageous when compared with lower energy density sources such as a battery. Fossil fuels are currently economical for decentralised energy use.

A (horizontal) drilling rig for natural gas.

Energy dependence on imported fossil fuels creates energy security risks for dependent countries. Oil dependence in particular has led to war, funding of radicals, monopolization, and socio-political instability.

Fossil fuels are non-renewable resources, which will eventually decline in production and become exhausted. While the processes that created fossil fuels are ongoing, fuels are consumed far more quickly than the natural rate of replenishment. Extracting fuels becomes increasingly costly as society consumes the most accessible fuel deposits. Extraction of fossil fuels results in environmental degradation, such as the strip mining and mountaintop removal of coal.

Fuel efficiency is a form of thermal efficiency, meaning the efficiency of a process that converts chemical potential energy contained in a carrier fuel into kinetic energy or work. The fuel economy is the energy efficiency of a particular vehicle, is given as a ratio of distance travelled per unit of fuel consumed. Weight-specific efficiency (efficiency per unit weight) may be stated for freight, and passenger-specific efficiency (vehicle efficiency per passenger). The inefficient atmospheric combustion (burning) of fossil fuels in vehicles, buildings, and power plants contributes to urban heat islands.

Conventional production of oil has peaked, conservatively, between 2007 and 2010. In 2010, it was estimated that an investment in non-renewable resources of $8 trillion would be required to maintain current levels of production for 25 years. In 2010, governments subsidized fossil fuels by an estimated $500 billion a year. Fossil fuels are also a source of greenhouse gas emissions, leading to concerns about global warming if consumption is not reduced.

The combustion of fossil fuels leads to the release of pollution into the atmosphere. The fossil fuels are mainly carbon compounds. During combustion, carbon dioxide is released, and also nitrogen oxides, soot and other fine particulates. Man-made carbon dioxide according to the IPCC contributes to global warming. Other emissions from fossil fuel power station include sulfur dioxide, carbon monoxide (CO), hydrocarbons, volatile organic compounds (VOC), mercury, arsenic, lead, cadmium, and other heavy metals including traces of uranium.

A typical coal plant generates billions of kilowatt hours per year.

Nuclear Fission

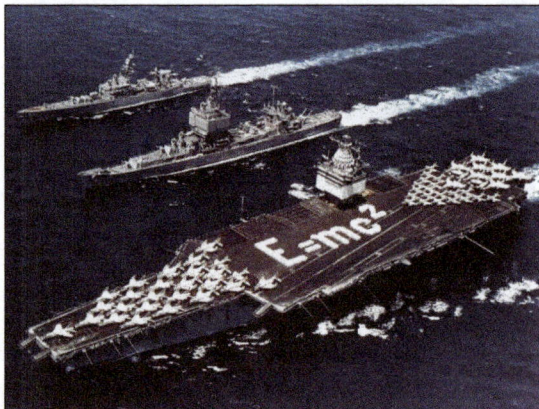

American nuclear powered ships, (top to bottom) cruisers USS *Bainbridge*, the USS *Long Beach* and the *USS Enterprise*, the longest ever naval vessel, and the first nuclear-powered aircraft carrier.

Picture above was taken in 1964 during a record setting voyage of 26,540 nmi (49,190 km) around the world in 65 days without refueling. Crew members are spelling out Einstein's mass-energy equivalence formula $E = mc^2$ on the flight deck.

The Russian nuclear-powered icebreaker NS Yamal on a joint scientific expedition with the NSF in 1994.

Nuclear power is the use of nuclear fission to generate useful heat and electricity. Fission of uranium produces nearly all economically significant nuclear power. Radioisotope thermoelectric generators form a very small component of energy generation, mostly in specialized applications such as deep space vehicles.

Nuclear power plants, excluding naval reactors, provided about 5.7% of the world's energy and 13% of the world's electricity in 2012.

In 2013, the IAEA report that there are 437 operational nuclear power reactors, in 31 countries, although not every reactor is producing electricity. In addition, there are approximately 140 naval vessels using nuclear propulsion in operation, powered by some 180 reactors. As of 2013, attaining a net energy gain from sustained nuclear fusion reactions, excluding natural fusion power sources such as the Sun, remains an ongoing area of international physics and engineering research. More than 60 years after the first attempts, commercial fusion power production remains unlikely before 2050.

There is an ongoing debate about nuclear power. Proponents, such as the World Nuclear Association, the IAEA and Environmentalists for Nuclear Energy contend that nuclear power is a safe, sustainable energy source that reduces carbon emissions. Opponents, such as Greenpeace International and NIRS, contend that nuclear power poses many threats to people and the environment.

Nuclear power plant accidents include the Chernobyl disaster, Fukushima Daiichi nuclear disaster, and the Three Mile Island accident. There have also been some nuclear submarine accidents. In terms of lives lost per unit of energy generated, analysis has determined that nuclear power has caused less fatalities per unit of energy generated than the other major sources of energy generation. Energy production from coal, petroleum, natural gas and hydropower has caused a greater number of fatalities per unit of energy generated due to air pollution and energy accident effects. However, the economic costs of nuclear power accidents is high, and meltdowns can take decades to clean up. The human costs of evacuations of affected populations and lost livelihoods is also significant.

Comparing Nuclear's *latent* cancer deaths, such as cancer with other energy sources *immediate* deaths per unit of energy generated(GWeyr). This study does not include fossil fuel related cancer and other indirect deaths created by the use of fossil fuel consumption in its "severe accident" classification, which would be an accident with more than 5 fatalities.

Nuclear power is a low carbon power generation method of producing electricity, with an analysis of the literature on its total life cycle emission intensity finding that it is similar to renewable sources in a comparison of greenhouse gas(GHG) emissions per unit of energy generated. Since the 1970s, nuclear fuel has displaced about 64 gigatonnes of carbon dioxide equivalent(GtCO2-eq) greenhouse gases, that would have otherwise resulted from the burning of oil, coal or natural gas in fossil-fuel power stations.

As of 2012, according to the IAEA, worldwide there were 68 civil nuclear power reactors under construction in 15 countries, approximately 28 of which in the People's Republic of China (PRC), with the most recent nuclear power reactor, as of May 2013, to be connected to the electrical grid, occurring on February 17, 2013 in Hongyanhe Nuclear Power Plant in the PRC. In the United States, two new Generation III reactors are under construction at Vogtle. U.S. nuclear industry officials expect five new reactors to enter service by 2020, all at existing plants. In 2013, four aging, uncompetitive, reactors were permanently closed.

Japan's 2011 Fukushima Daiichi nuclear accident, which occurred in a reactor design from the 1960s, prompted a rethink of nuclear safety and nuclear energy policy in many countries. Germany decided to close all its reactors by 2022, and Italy has banned nuclear power. Following Fukushima, in 2011 the International Energy Agency halved its estimate of additional nuclear generating capacity to be built by 2035.

Recent experiments in extraction of uranium use polymer ropes that are coated with a substance that selectively absorbs uranium from seawater. This process could make the considerable volume of uranium dissolved in seawater exploitable for energy production. Since ongoing geologic processes carry uranium to the sea in amounts comparable to the amount that would be extracted by this process, in a sense the sea-borne uranium becomes a sustainable resource.

Fission Economics

Fukushima Daiichi nuclear disaster.

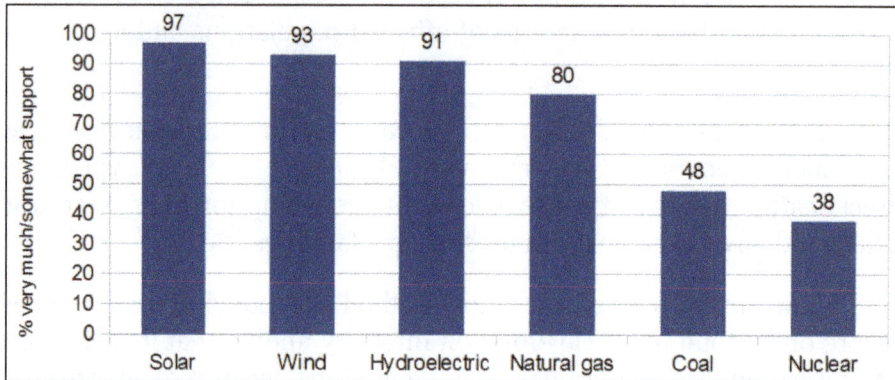

Low global public support for nuclear fission in the aftermath of Fukushima.

The economics of new nuclear power plants is a controversial subject, since there are diverging views on this topic, and multibillion-dollar investments ride on the choice of an energy source. Nuclear power plants typically have high capital costs for building the plant, but low direct fuel costs. In recent years there has been a slowdown of electricity demand growth and financing has become more difficult, which affects large projects such as nuclear reactors, with very large upfront costs and long project cycles which carry a large variety of risks. In Eastern Europe, a number of long-established projects are struggling to find finance, notably Belene in Bulgaria and the additional reactors at Cernavoda in Romania, and some potential backers have pulled out. Where cheap gas is available and its future supply relatively secure, this also poses a major problem for nuclear projects.

Analysis of the economics of nuclear power must take into account who bears the risks of future uncertainties. To date all operating nuclear power plants were developed by state-owned or regulated utility monopolies where many of the risks associated with construction costs, operating performance, fuel price, and other factors were borne by consumers rather than suppliers. Many countries have now liberalized the electricity market where these risks, and the risk of cheaper competitors emerging before capital costs are recovered, are borne by plant suppliers and operators rather than consumers, which leads to a significantly different evaluation of the economics of new nuclear power plants.

Fukushima

Following the 2011 Fukushima Daiichi nuclear disaster – the second worst nuclear incident, that displaced 50,000 households after radioactive material leaked into the air, soil and sea, and with subsequent radiation checks leading to bans on some shipments of vegetables and fish – a global public support survey by Ipsos for energy sources was published and nuclear fission was found to be the least popular

Costs

Costs are likely to go up for currently operating and new nuclear power plants, due to increased requirements for on-site spent fuel management and elevated design basis threats. While first of their kind designs, such as the EPRs under construction are behind schedule and over-budget, of the seven South Korean APR-1400s presently under construction worldwide, two are in S.Korea at the Hanul Nuclear Power Plant and four are at the largest nuclear station construction project in the world as of 2016, in the United Arab Emirates at the planned Barakah nuclear power plant. The first reactor, Barakah-1 is 85% completed and on schedule for grid-connection during 2017. Two of the four EPRs under construction (in Finland and France) are significantly behind schedule and substantially over cost.

Renewable Sources

Wind, sun, and hydroelectricity are three renewable energy sources.

Renewable energy is generally defined as energy that comes from resources which are naturally replenished on a human timescale such as sunlight, wind, rain, tides, waves and geothermal heat.

Renewable energy replaces conventional fuels in four distinct areas: electricity generation, hot water/space heating, motor fuels, and rural (off-grid) energy services.

About 16% of global final energy consumption presently comes from renewable resources, with 10% of all energy from traditional biomass, mainly used for heating, and 3.4% from hydroelectricity. New renewables (small hydro, modern biomass, wind, solar, geothermal, and biofuels) account for another 3% and are growing rapidly. At the national level, at least 30 nations around the world already have renewable energy contributing more than 20% of energy supply. National renewable energy markets are projected to continue to grow strongly in the coming decade and beyond. Wind power, for example, is growing at the rate of 30% annually, with a worldwide installed capacity of 282,482 megawatts (MW) at the end of 2012.

Renewable energy resources exist over wide geographical areas, in contrast to other energy sources, which are concentrated in a limited number of countries. Rapid deployment of renewable energy and energy efficiency is resulting in significant energy security, climate change mitigation, and economic benefits. In international public opinion surveys there is strong support for promoting renewable sources such as solar power and wind power.

While many renewable energy projects are large-scale, renewable technologies are also suited to rural and remote areas and developing countries, where energy is often crucial in human development. United Nations' Secretary-General Ban Ki-moon has said that renewable energy has the ability to lift the poorest nations to new levels of prosperity.

Hydroelectricity

The 22,500 MW Three Gorges Dam in China – The world's largest hydroelectric power station.

Hydroelectricity is electric power generated by hydropower; the force of falling or flowing water. In 2015 hydropower generated 16.6% of the world's total electricity and 70% of all renewable electricity and is expected to increase about 3.1% each year for the next 25 years.

Hydropower is produced in 150 countries, with the Asia-Pacific region generating 32 percent of global hydropower in 2010. China is the largest hydroelectricity producer, with 721 terawatt-hours of production in 2010, representing around 17 percent of domestic electricity use. There are now three hydroelectricity plants larger than 10 GW: the Three Gorges Dam in China, Itaipu Dam across the Brazil/Paraguay border, and Guri Dam in Venezuela.

The cost of hydroelectricity is relatively low, making it a competitive source of renewable electricity. The average cost of electricity from a hydro plant larger than 10 megawatts is 3 to 5 U.S. cents per kilowatt-hour. Hydro is also a flexible source of electricity since plants can be ramped up and down very quickly to adapt to changing energy demands. However, damming interrupts the flow of rivers and can harm local ecosystems, and building large dams and reservoirs often involves displacing people and wildlife. Once a hydroelectric complex is constructed, the project produces no direct waste, and has a considerably lower output level of the greenhouse gas carbon dioxide than fossil fuel powered energy plants.

Wind

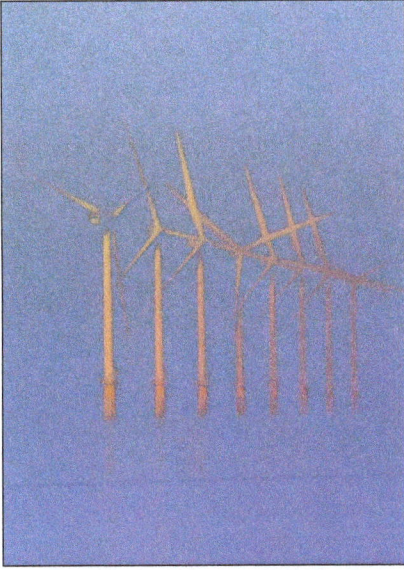

Burbo Bank Offshore Wind Farm.

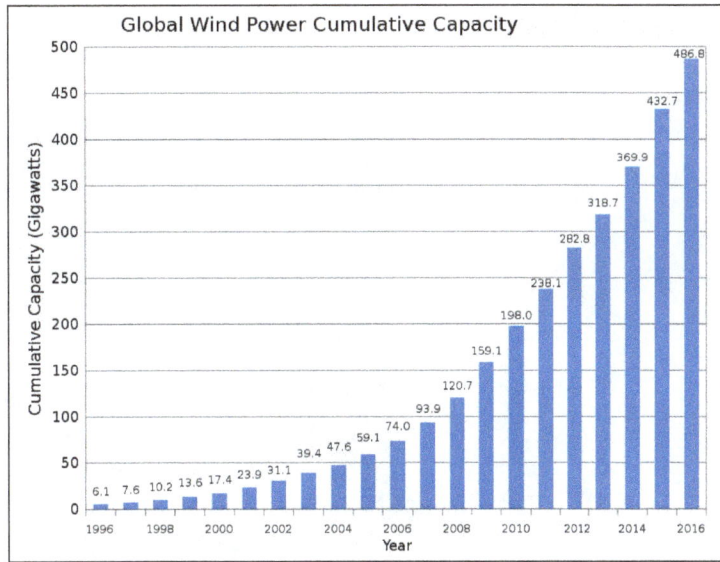

Global growth of wind power capacity.

Wind power harnesses the power of the wind to propel the blades of wind turbines. These turbines cause the rotation of magnets, which creates electricity. Wind towers are usually built together on wind farms. There are offshore and onshore wind farms. Global wind power capacity has expanded rapidly to 336 GW in June 2014, and wind energy production was around 4% of total worldwide electricity usage, and growing rapidly.

Wind power is widely used in Europe, Asia, and the United States. Several countries have achieved relatively high levels of wind power penetration, such as 21% of stationary electricity production in Denmark, 18% in Portugal, 16% in Spain, 14% in Ireland, and 9% in Germany in 2010. By 2011, at times over 50% of electricity in Germany and Spain came from wind and solar power. As of 2011, 83 countries around the world are using wind power on a commercial basis.

Many of the world's largest onshore wind farms are located in the United States, China, and India. Most of the world's largest offshore wind farms are located in Denmark, Germany and the United Kingdom. The two largest offshore wind farm are currently the 630 MW London Array and Gwynt y Môr.

Large onshore wind farms		
Wind farm	Current capacity (M W)	Country
Alta (Oak Creek-Mojave)	1,320	USA
Jaisalmer Wind Park	1,064	India
Roscoe Wind Farm	781	USA
Horse Hollow Wind Energy Center	735	USA
Capricorn Ridge Wind Farm	662	USA
Fântânele-Cogealac Wind Farm	600	Romania
Fowler Ridge Wind Farm	599	USA

Solar

Part of the 354 MW SEGS solar complex in northern San Bernardino County.

The 150 MW Andasol Solar Power Station is a concentrated solar power plant.

Solar energy, radiant light and heat from the sun, is harnessed using a range of ever-evolving technologies such as solar heating, solar photovoltaics, solar thermal electricity, solar architecture and artificial photosynthesis.

Solar technologies are broadly characterized as either passive solar or active solar depending on the way they capture, convert and distribute solar energy. Active solar techniques include the use of photovoltaic panels and solar thermal collectors to harness the energy. Passive solar techniques include orienting a building to the Sun, selecting materials with favorable thermal mass or light dispersing properties, and designing spaces that naturally circulate air.

In 2011, the International Energy Agency said that "the development of affordable, inexhaustible and clean solar energy technologies will have huge longer-term benefits. It will increase countries' energy security through reliance on an indigenous, inexhaustible and mostly import-independent resource, enhance sustainability, reduce pollution, lower the costs of mitigating climate change, and keep fossil fuel prices lower than otherwise. These advantages are global. Hence the additional costs of the incentives for early deployment should be considered learning investments; they must be wisely spent and need to be widely shared". More than 100 countries use solar PV.

The Topaz Solar Farm is one of the world's largest solar power stations.

Photovoltaics (PV) is a method of generating electrical power by converting solar radiation into direct current electricity using semiconductors that exhibit the photovoltaic effect. Photovoltaic power generation employs solar panels composed of a number of solar cells containing a photovoltaic material. Materials presently used for photovoltaics include monocrystalline silicon, polycrystalline silicon, amorphous silicon, cadmium telluride, and copper indium gallium selenide/sulfide. Due to the increased demand for renewable energy sources, the manufacturing of solar cells and photovoltaic arrays has advanced considerably in recent years.

Solar photovoltaics is a sustainable energy source. By the end of 2018, a total of 505 GW had been installed worldwide with 100 GW installed in that year.

Driven by advances in technology and increases in manufacturing scale and sophistication, the cost of photovoltaics has declined steadily since the first solar cells were manufactured, and the levelised cost of electricity (LCOE) from PV is competitive with conventional electricity sources in an expanding list of geographic regions. Net metering and financial incentives, such as preferential feed-in tariffs for solar-generated electricity, have supported solar PV installations in many countries. The Energy Payback Time (EPBT), also known as *energy amortization*, depends on the location's annual solar insolation and temperature profile, as well as on the used type of PV-technology. For conventional crystalline silicon photovoltaics, the EPBT is higher than for thin-film technologies such as CdTe-PV or CPV-systems. Moreover, the payback time decreased in the recent years due to a number of improvements such as solar cell efficiency and more economic manufacturing processes. As of 2014, photovoltaics recoup on average the energy needed to manufacture them in 0.7 to 2 years. This results in about 95% of net-clean energy produced by a solar rooftop PV system over a 30-year life-time. Installations may be ground-mounted (and sometimes integrated with farming and grazing) or built into the roof or walls of a building (either building-integrated photovoltaics or simply rooftop).

Biofuels

A bus fueled by biodiesel.

Information on pump regarding ethanol fuel blend up to 10%, California.

A biofuel is a fuel that contains energy from geologically recent carbon fixation. These fuels are produced from living organisms. Examples of this carbon fixation occur in plants and microalgae. These fuels are made by a biomass conversion (biomass refers to recently living organisms, most often referring to plants or plant-derived materials). This biomass can be converted to convenient energy containing substances in three different ways: thermal conversion, chemical conversion, and biochemical conversion. This biomass conversion can result in fuel in solid, liquid, or gas form. This new biomass can be used for biofuels. Biofuels have increased in popularity because of rising oil prices and the need for energy security.

Bioethanol is an alcohol made by fermentation, mostly from carbohydrates produced in sugar or starch crops such as corn or sugarcane. Cellulosic biomass, derived from non-food sources, such as trees and grasses, is also being developed as a feedstock for ethanol production. Ethanol can be used as a fuel for vehicles in its pure form, but it is usually used as a gasoline additive to increase octane and improve vehicle emissions. Bioethanol is widely used in the USA and in Brazil. Current plant design does not provide for converting the lignin portion of plant raw materials to fuel components by fermentation.

Biodiesel is made from vegetable oils and animal fats. Biodiesel can be used as a fuel for vehicles in its pure form, but it is usually used as a diesel additive to reduce levels of particulates, carbon monoxide, and hydrocarbons from diesel-powered vehicles. Biodiesel is produced from oils or fats using transesterification and is the most common biofuel in Europe. However, research is underway on producing renewable fuels from decarboxylation.

In 2010, worldwide biofuel production reached 105 billion liters (28 billion gallons US), up 17% from 2009, and biofuels provided 2.7% of the world's fuels for road transport, a contribution largely made up of ethanol and biodiesel. Global ethanol fuel production reached 86 billion liters (23 billion gallons US) in 2010, with the United States and Brazil as the world's top producers, accounting together for 90% of global production. The world's largest biodiesel producer is the European Union, accounting for 53% of all biodiesel production in 2010. As of 2011, mandates for blending biofuels exist in 31 countries at the national level and in 29 states or provinces. The International Energy Agency has a goal for biofuels to meet more than a quarter of world demand for transportation fuels by 2050 to reduce dependence on petroleum and coal.

Geothermal

Steam rising from the Nesjavellir Geothermal Power Station in Iceland.

Geothermal energy is thermal energy generated and stored in the Earth. Thermal energy is the energy that determines the temperature of matter. The geothermal energy of the Earth's crust originates from the original formation of the planet (20%) and from radioactive decay of minerals (80%). The geothermal gradient, which is the difference in temperature between the core of the planet and its surface, drives a continuous conduction of thermal energy in the form of heat from the core to the surface.

Earth's internal heat is thermal energy generated from radioactive decay and continual heat loss from Earth's formation. Temperatures at the core-mantle boundary may reach over 4000 °C (7,200 °F). The high temperature and pressure in Earth's interior cause some rock to melt and solid mantle to behave plastically, resulting in portions of mantle convecting upward since it is lighter than the surrounding rock. Rock and water is heated in the crust, sometimes up to 370 °C (700 °F).

From hot springs, geothermal energy has been used for bathing since Paleolithic times and for space heating since ancient Roman times, but it is now better known for electricity generation. Worldwide, 11,400 megawatts (MW) of geothermal power is online in 24 countries in 2012. An additional 28 gigawatts of direct geothermal heating capacity is installed for district heating, space heating, spas, industrial processes, desalination and agricultural applications in 2010.

Geothermal power is cost effective, reliable, sustainable, and environmentally friendly, but has historically been limited to areas near tectonic plate boundaries. Recent technological advances have dramatically expanded the range and size of viable resources, especially for applications such as home heating, opening a potential for widespread exploitation. Geothermal wells release greenhouse gases trapped deep within the earth, but these emissions are much lower per energy unit than those of fossil fuels. As a result, geothermal power has the potential to help mitigate global warming if widely deployed in place of fossil fuels.

The Earth's geothermal resources are theoretically more than adequate to supply humanity's energy needs, but only a very small fraction may be profitably exploited. Drilling and exploration for deep resources is very expensive. Forecasts for the future of geothermal power depend on assumptions about technology, energy prices, subsidies, and interest rates. Pilot programs like EWEB's customer opt in Green Power Program show that customers would be willing to pay a little more for a renewable energy source like geothermal. But as a result of government assisted research and industry experience, the cost of generating geothermal power has decreased by 25% over the past two decades. In 2001, geothermal energy cost between two and ten US cents per kWh.

Oceanic

Marine energy or marine power (also sometimes referred to as ocean energy, ocean power, or marine and hydrokinetic energy) refers to the energy carried by ocean waves, tides, salinity, and ocean temperature differences. The movement of water in the world's oceans creates a vast store of kinetic energy, or energy in motion. This energy can be harnessed to generate electricity to power homes, transport and industries.

The term marine energy encompasses both wave power i.e. power from surface waves, and tidal power i.e. obtained from the kinetic energy of large bodies of moving water. Offshore wind power is not a form of marine energy, as wind power is derived from the wind, even if the wind turbines are placed over water. The oceans have a tremendous amount of energy and are close to many if not most concentrated populations. Ocean energy has the potential of providing a substantial amount of new renewable energy around the world.

100% Renewable Energy

The incentive to use 100% renewable energy, for electricity, transport, or even total primary energy supply globally, has been motivated by global warming and other ecological as well as economic concerns. Renewable energy use has grown much faster than anyone anticipated. The Intergovernmental Panel on Climate Change has said that there are few fundamental technological limits to integrating a portfolio of renewable energy technologies to meet most of total global energy demand. At the national level, at least 30 nations around the world already have renewable energy contributing more than 20% of energy supply. Also, Professors S. Pacala and Robert H. Socolow have developed a series of "stabilization wedges" that can allow us to maintain our quality of life while avoiding catastrophic climate change, and "renewable energy sources," in aggregate, constitute the largest number of their "wedges."

Mark Z. Jacobson says producing all new energy with wind power, solar power, and hydropower by 2030 is feasible and existing energy supply arrangements could be replaced by 2050. Barriers to implementing the renewable energy plan are seen to be "primarily social and political, not technological or economic". Jacobson says that energy costs with a wind, solar, water system should be similar to today's energy costs.

Similarly, in the United States, the independent National Research Council has noted that "sufficient domestic renewable resources exist to allow renewable electricity to play a significant role in future electricity generation and thus help confront issues related to climate change, energy security, and the escalation of energy costs. Renewable energy is an attractive option because renewable resources available in the United States, taken collectively, can supply significantly greater amounts of electricity than the total current or projected domestic demand."

Critics of the "100% renewable energy" approach include Vaclav Smil and James E. Hansen. Smil and Hansen are concerned about the variable output of solar and wind power, but Amory Lovins argues that the electricity grid can cope, just as it routinely backs up nonworking coal-fired and nuclear plants with working ones.

Google spent $30 million on their RE<C project to develop renewable energy and stave off catastrophic climate change. The project was cancelled after concluding that a best-case scenario for rapid advances in renewable energy could only result in emissions 55 percent below the fossil fuel projections for 2050.

Increased Energy Efficiency

A spiral-type integrated compact fluorescent lamp, which has been popular among North American consumers since its introduction in the mid-1990s.

Although increasing the efficiency of energy use is not energy development per se, it may be considered under the topic of energy development since it makes existing energy sources available to do work.

Efficient energy use reduces the amount of energy required to provide products and services. For example, insulating a home allows a building to use less heating and cooling energy to maintain a comfortable temperature. Installing fluorescent lamps or natural skylights reduces the amount of energy required for illumination compared to incandescent light bulbs. Compact fluorescent lights use two-thirds less energy and may last 6 to 10 times longer than incandescent lights. Improvements in energy efficiency are most often achieved by adopting an efficient technology or production process.

Reducing energy use may save consumers money, if the energy savings offsets the cost of an energy efficient technology. Reducing energy use reduces emissions. According to the International Energy Agency, improved energy efficiency in buildings, industrial processes and transportation could reduce the world's energy needs in 2050 by one third, and help control global emissions of greenhouse gases.

Energy efficiency and renewable energy are said to be the *twin pillars* of sustainable energy policy. In many countries energy efficiency is also seen to have a national security benefit because it can be used to reduce the level of energy imports from foreign countries and may slow down the rate at which domestic energy resources are depleted.

It's been discovered "that for OECD countries, wind, geothermal, hydro and nuclear have the lowest hazard rates among energy sources in production".

Transmission

An elevated section of the Alaska Pipeline.

While new sources of energy are only rarely discovered or made possible by new technology, distribution technology continually evolves. The use of fuel cells in cars, for example, is an anticipated delivery technology. This topic presents the various delivery technologies that have been important to historic energy development.

Shipping and Pipelines

Coal, petroleum and their derivatives are delivered by boat, rail, or road. Petroleum and natural gas may also be delivered by pipeline, and coal via a Slurry pipeline. Fuels such as gasoline and LPG may also be delivered via aircraft. Natural gas pipelines must maintain a certain minimum pressure to function correctly. The higher costs of ethanol transportation and storage are often prohibitive.

Wired Energy Transfer

Electrical grid – pylons and Cables Distribute Power.

Electricity grids are the networks used to transmit and distribute power from production source to end user, when the two may be hundreds of kilometres away. Sources include electrical generation plants such as a nuclear reactor, coal burning power plant, etc. A combination of sub-stations and transmission lines are used to maintain a constant flow of electricity. Grids may suffer from transient blackouts and brownouts, often due to weather damage. During certain extreme space weather events solar wind can interfere with transmissions. Grids also have a predefined carrying capacity or load that cannot safely be exceeded. When power requirements exceed what's available, failures are inevitable. To prevent problems, power is then rationed.

Industrialised countries such as Canada, the US, and Australia are among the highest per capita consumers of electricity in the world, which is possible thanks to a widespread electrical distribution network. The US grid is one of the most advanced, although infrastructure maintenance is becoming a problem. Current Energy provides a realtime overview of the electricity supply and demand for California, Texas, and the Northeast of the US. African countries with small scale electrical grids have a correspondingly low annual per capita usage of electricity. One of the most powerful power grids in the world supplies power to the state of Queensland, Australia.

Wireless Energy Transfer

Wireless power transfer is a process whereby electrical energy is transmitted from a power source to an electrical load that does not have a built-in power source, without the use of interconnecting wires. Currently available technology is limited to short distances and relatively low power level.

Orbiting solar power collectors would require wireless transmission of power to Earth. The proposed method involves creating a large beam of microwave-frequency radio waves, which would be aimed at a collector antenna site on the Earth. Formidable technical challenges exist to ensure the safety and profitability of such a scheme.

Storage

The Ffestiniog Power Station in Wales, United Kingdom. Pumped-storage
hydroelectricity (PSH) is used for grid energy storage.

Energy storage is accomplished by devices or physical media that store energy to perform useful operation at a later time. A device that stores energy is sometimes called an accumulator.

All forms of energy are either potential energy (e.g. Chemical, gravitational, electrical energy, temperature differential, latent heat, etc.) or kinetic energy (e.g. momentum). Some technologies provide only short-term energy storage, and others can be very long-term such as power to gas using hydrogen or methane and the storage of heat or cold between opposing seasons in deep aquifers or bedrock. A wind-up clock stores potential energy (in this case mechanical, in the spring tension), a battery stores readily convertible chemical energy to operate a mobile phone, and a hydroelectric dam stores energy in a reservoir as gravitational potential energy. Ice storage tanks store ice (thermal energy in the form of latent heat) at night to meet peak demand for cooling. Fossil fuels such as coal and gasoline store ancient energy derived from sunlight by organisms that later died, became buried and over time were then converted into these fuels. Even food (which is made by the same process as fossil fuels) is a form of energy stored in chemical form.

Energy generators past and present at Doel, Belgium: 17th-century windmill Scheldemolen and 20th-century Doel Nuclear Power Station.

Since prehistory, when humanity discovered fire to warm up and roast food, through the Middle Ages in which populations built windmills to grind the wheat, until the modern era in which nations can get electricity splitting the atom. Man has sought endlessly for energy sources.

Except nuclear, geothermal and tidal, all other energy sources are from current solar isolation or from fossil remains of plant and animal life that relied upon sunlight. Ultimately, solar energy itself is the result of the Sun's nuclear fusion. Geothermal power from hot, hardened rock above the magma of the Earth's core is the result of the decay of radioactive materials present beneath the Earth's crust, and nuclear fission relies on man-made fission of heavy radioactive elements in the Earth's crust; in both cases these elements were produced in supernova explosions before the formation of the solar system.

Since the beginning of the Industrial Revolution, the question of the future of energy supplies has been of interest. In 1865, William Stanley Jevons published *The Coal Question* in which he saw that the reserves of coal were being depleted and that oil was an ineffective replacement. In 1914, U.S. Bureau of Mines stated that the total production was 5.7 billion barrels (910,000,000 m³). In 1956, Geophysicist M. King Hubbert deduces that U.S. oil production would peak between 1965 and 1970 and that oil production will peak "within half a century" on the basis of 1956 data. In 1989, predicted peak by Colin Campbell In 2004, OPEC estimated, with substantial investments, it would nearly double oil output by 2025.

Sustainability

The environmental movement has emphasized sustainability of energy use and development.

Renewable energy is sustainable in its production; the available supply will not be diminished for the foreseeable future - millions or billions of years. "Sustainability" also refers to the ability of the environment to cope with waste products, especially air pollution. Sources which have no direct waste products (such as wind, solar, and hydropower) are brought up on this point. With global demand for energy growing, the need to adopt various energy sources is growing. Energy conservation is an alternative or complementary process to energy development. It reduces the demand for energy by using it efficiently.

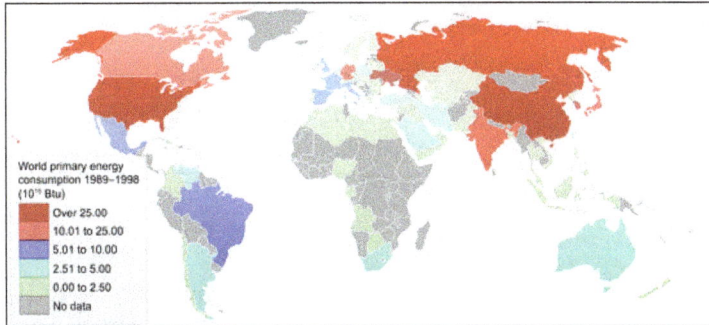

Energy consumption from 1989 to 1999.

Resilience

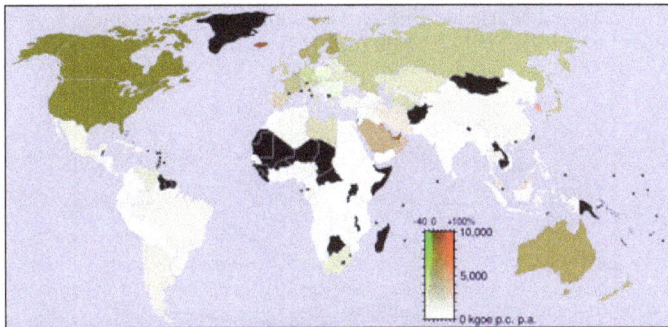

Energy consumption *per capita* (2001). Red hues indicate increase, green hues decrease of consumption during the 1990s.

Some observers contend that idea of "energy independence" is an unrealistic and opaque concept. The alternative offer of "energy resilience" is a goal aligned with economic, security, and energy realities. It was argued that simply switching to domestic energy would not be secure inherently because the true weakness is the interdependent and vulnerable energy infrastructure of the United States. Key aspects such as gas lines and the electrical power grid are centralized and easily susceptible to disruption. They conclude that a "resilient energy supply" is necessary for both national security and the environment. They recommend a focus on energy efficiency and renewable energy that is decentralized.

In 2008, former Intel Corporation Chairman and CEO Andrew Grove looked to energy resilience, arguing that complete independence is unfeasible given the global market for energy. He describes energy resilience as the ability to adjust to interruptions in the supply of energy. To that end, he suggests the U.S. make greater use of electricity. Electricity can be produced from a variety of sources. A diverse energy supply will be less affected by the disruption in supply of any one source. He reasons that another feature of electrification is that electricity is "sticky" – meaning the electricity produced in the U.S. is to stay there because it cannot be transported overseas. According to Grove, a key aspect of advancing electrification and energy resilience will be converting the U.S.

automotive fleet from gasoline-powered to electric-powered. This, in turn, will require the modernization and expansion of the electrical power grid. As organizations such as The Reform Institute have pointed out, advancements associated with the developing smart grid would facilitate the o it to charge their batteries.

Present and Future

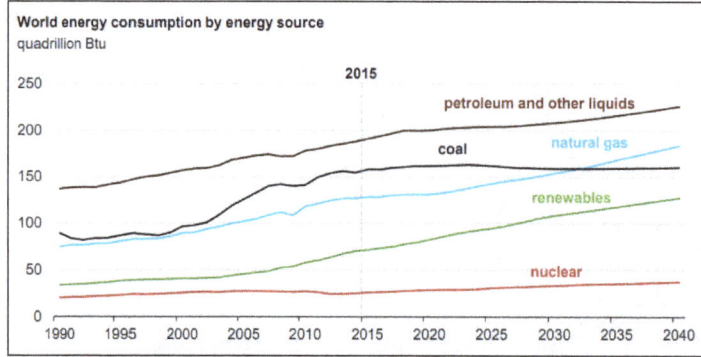

Outlook—World Energy Consumption by Fuel (as of 2011).
Liquid fuels incl. Biofuels Coal Natural Gas Renewable fuels Nuclear fuels.

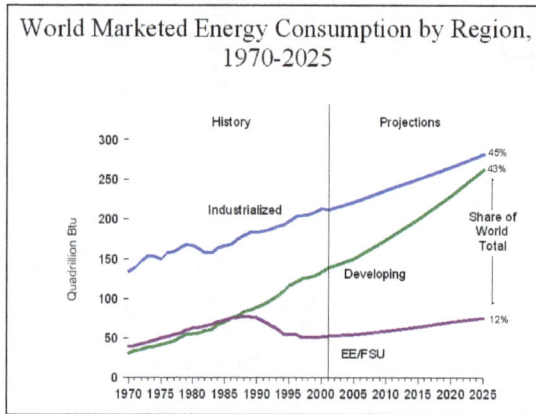

Increasing share of energy consumption by developing nations.
Industrialized nations Developing nations EE/Former Soviet Union.

Extrapolations from current knowledge to the future offer a choice of energy futures. Predictions parallel the Malthusian catastrophe hypothesis. Numerous are complex models based scenarios as pioneered by *Limits to Growth*. Modeling approaches offer ways to analyze diverse strategies, and hopefully find a road to rapid and sustainable development of humanity. Short term energy crises are also a concern of energy development. Extrapolations lack plausibility, particularly when they predict a continual increase in oil consumption.

Energy production usually requires an energy investment. Drilling for oil or building a wind power plant requires energy. The fossil fuel resources that are left are often increasingly difficult to extract and convert. They may thus require increasingly higher energy investments. If investment is greater than the value of the energy produced by the resource, it is no longer an effective energy source. These resources are no longer an energy source but may be exploited for value as raw materials. New technology may lower the energy investment required to extract and convert the resources, although ultimately basic physics sets limits that cannot be exceeded.

Between 1950 and 1984, as the Green Revolution transformed agriculture around the globe, world grain production increased by 250%. The energy for the Green Revolution was provided by fossil fuels in the form of fertilizers (natural gas), pesticides (oil), and hydrocarbon fueled irrigation. The peaking of world hydrocarbon production (peak oil) may lead to significant changes, and require sustainable methods of production. One vision of a sustainable energy future involves all human structures on the earth's surface (i.e., buildings, vehicles and roads) doing artificial photosynthesis (using sunlight to split water as a source of hydrogen and absorbing carbon dioxide to make fertilizer) efficiently than plants.

With contemporary space industry's economic activity and the related private spaceflight, with the manufacturing industries, that go into Earth's orbit or beyond, delivering them to those regions will require further energy development. Researchers have contemplated space-based solar power for collecting solar power for use on Earth. Space-based solar power has been in research since the early 1970s. Space-based solar power would require construction of collector structures in space. The advantage over ground-based solar power is higher intensity of light, and no weather to interrupt power collection.

Energy Recovery

Energy recovery includes any technique or method of minimizing the input of energy to an overall system by the exchange of energy from one sub-system of the overall system with another. The energy can be in any form in either subsystem, but most energy recovery systems exchange thermal energy in either sensible or latent form.

In some circumstances the use of an enabling technology, either diurnal thermal energy storage or seasonal thermal energy storage (STES, which allows heat or cold storage between opposing seasons), is necessary to make energy recovery practicable. One example is waste heat from air conditioning machinery stored in a buffer tank to aid in night time heating. Another is an STES application at a foundry in Sweden. Waste heat is recovered and stored in a large mass of native bedrock which is penetrated by a cluster of 140 heat exchanger equipped boreholes (155mm diameter) that are 150m deep. This store is used for heating an adjacent factory as needed, even months later. An example of using STES to recover and utilize natural heat that otherwise would be wasted is the Drake Landing Solar Community in Alberta, Canada. The community uses a cluster of boreholes in bedrock for interseasonal heat storage, and this enables obtaining 97 percent of the year-round space heating from solar thermal collectors on the garage roofs. Another STES application is recovering the cold of winter by circulating water through a dry cooling tower, and using that to chill a deep aquifer or borehole cluster. The chill is later recovered from the storage for summer air conditioning. With a coefficient of performance (COP) of 20 to 40, this method of cooling can be ten times more efficient than conventional air conditioning.

Principle

A common application of this principle is in systems which have an *exhaust stream* or *waste stream* which is transferred from the system to its surroundings. Some of the energy in that flow of material (often gaseous or liquid) may be transferred to the *make-up* or *input* material flow.

This *input* mass flow often comes from the system's surroundings, which, being at ambient conditions, are at a lower temperature than the *waste stream*. This temperature differential allows heat transfer and thus energy transfer, or in this case, recovery. Thermal energy is often recovered from liquid or gaseous waste streams to *fresh make-up* air and water intakes in buildings, such as for the HVAC systems, or process systems.

System Approach

Energy consumption is a key part of most human activities. This consumption involves converting one energy system to another, for example: The conversion of mechanical energy to electrical energy, which can then power computers, light, motors etc. The input energy propels the work and is mostly converted to heat or follows the product in the process as output energy. Energy recovery systems harvest the output power and provide this as input power to the same or another process.

An energy recovery system will close this energy cycle to prevent the input power from being released back to nature and rather be used in other forms of desired work.

Examples of Energy Recovery

Heat recovery is implemented in heat sources like e.g. a steel mill. Heated cooling water from the process is sold for heating of homes, shops and offices in the surrounding area.

- Regenerative braking is used in electric cars, trains, heavy cranes etc. where the energy consumed when elevating the potential is returned to the electric supplier when released.

- Active pressure reduction systems where the differential pressure in a pressurized fluid flow is recovered rather than converted to heat in a pressure reduction valve and released.

- Energy recovery ventilation.

- Energy recycling.

- Water heat recycling.

- Heat recovery ventilation.

- Heat recovery steam generator.

- Cyclone Waste Heat Engine.

- Hydrogen turboexpander-generator.

- Thermal diode.

- Thermal oxidizer.

- Thermoelectric Modules.

- Waste heat recovery units.

Electric Turbo Compound (ETC)

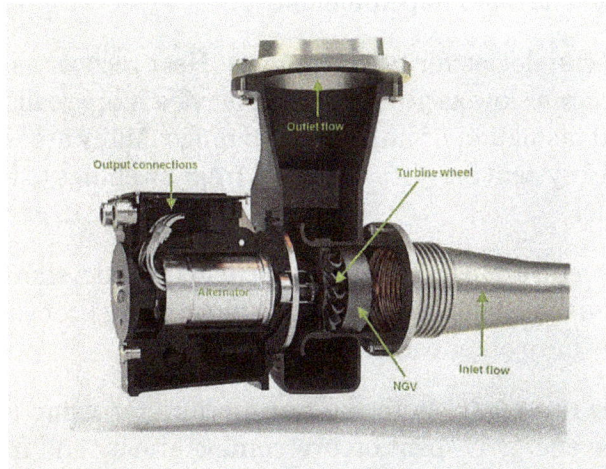

Electric Turbo Compound (ETC) cross section.

Electric Turbo Compounding (ETC) is a technology solution to the challenge of improving the fuel efficiency of gas and diesel engines by recovering waste energy from the exhaust gases.

Environmental Impact

There is a large potential for energy recovery in compact systems like large industries and utilities. Together with energy conservation, it should be possible to dramatically reduce world energy consumption. The effect of this will then be:

- Reduced number of coal-fired power plants.

- Reduced airborne particles, NO_x and CO_2 – improved air quality.

- Slowing or reducing climate change.

- Lower fuel bills on transport.

- Longer availability of crude oil.

- Change of industries and economies not fully researched.

In 2008 Tom Casten, chairman of Recycled Energy Development, said that "We think we could make about 19 to 20 percent of U.S. electricity with heat that is currently thrown away by industry."

A 2007 Department of Energy study found the potential for 135,000 megawatts of combined heat and power (which uses energy recovery) in the U.S., and a Lawrence Berkley National Laboratory study identified about 64,000 megawatts that could be obtained from industrial waste energy, not counting CHP. These studies suggest that about 200,000 megawatts, or 20%, of total power capacity could come from energy recycling in the U.S. Widespread use of energy recycling could therefore reduce global warming emissions by an estimated 20 percent. Indeed, as of 2005, about 42% of U.S. greenhouse gas pollution came from the production of electricity and 27% from the production of heat.

It is, however, difficult to quantify the environmental impact of a global energy recovery implementation in some sectors. The main impediments are:

- Lack of efficient technologies for private homes. Heat recovery systems in private homes can have an efficiency as low as 30% or less. It may be more realistic to use energy conservation like thermal insulation or improved buildings. Many areas are more dependent on forced cooling and a system for extracting heat from dwellings to be used for other uses are not widely available.

- Ineffective infrastructure. Heat recovery in particular need a short distance from producer to consumer to be viable. A solution may be to move a large consumer to the vicinity of the producer. This may have other complications.

- Transport sector is not ready. With the transport sector using about 20% of the energy supply, most of the energy is spent on overcoming gravity and friction. Electric cars with regenerative braking seem to be the best candidate for energy recovery. Wind systems on ships are under development. Very little work on the airline industry is known in this field.

Energy Audit

An energy audit is an inspection survey and an analysis of energy flows for energy conservation in a building. It may include a process or system to reduce the amount of energy input into the system without negatively affecting the output. In commercial and industrial real estate, an energy audit is the first step in identifying opportunities to reduce energy expense and carbon footprint.

Principle

When the object of study is an occupied building then reducing energy consumption while maintaining or improving human comfort, health and safety are of primary concern. Beyond simply identifying the sources of energy use, an energy audit seeks to prioritize the energy uses according to the greatest to least cost effective opportunities for energy savings.

Home Energy Audit

A *home energy audit* is a service where the energy efficiency of a house is evaluated by a person using professional equipment (such as blower doors and infrared cameras), with the aim to suggest the best ways to improve energy efficiency in heating and cooling the house.

An energy audit of a home may involve recording various characteristics of the building envelope including the walls, ceilings, floors, doors, windows, and skylights. For each of these components the area and resistance to heat flow (R-value) is measured or estimated. The leakage rate or infiltration of air through the building envelope is of concern, both of which are strongly affected by window construction and quality of door seals such as weatherstripping. The goal of this exercise is to quantify the building's overall thermal performance. The audit may also assess the efficiency, physical condition, and programming of mechanical systems such as the heating, ventilation, air conditioning equipment, and thermostat.

A home energy audit may include a written report estimating energy use given local climate criteria, thermostat settings, roof overhang, and solar orientation. This could show energy use for a given time period, say a year, and the impact of any suggested improvements per year. The accuracy of energy estimates are greatly improved when the homeowner's billing history is available showing the quantities of electricity, natural gas, fuel oil, or other energy sources consumed over a one or two-year period.

Some of the greatest effects on energy use are user behavior, climate, and age of the home. An energy audit may therefore include an interview of the homeowners to understand their patterns of use over time. The energy billing history from the local utility company can be calibrated using heating degree day and cooling degree day data obtained from recent, local weather data in combination with the thermal energy model of the building. Advances in computer-based thermal modeling can take into account many variables affecting energy use.

A home energy audit is often used to identify cost effective ways to improve the comfort and efficiency of buildings. In addition, homes may qualify for energy efficiency grants from central government.

Recently, the improvement of smartphone technology has enabled homeowners to perform relatively sophisticated energy audits of their own homes. This technique has been identified as a method to accelerate energy efficiency improvements.

United States

In the United States, this kind of service can often be facilitated by:

- Public utility companies, or their energy conservation department.

- Independent, private-sector companies such as energy services company, insulation contractor, or air sealing specialist.

- (US) State energy office.

Utility companies may provide this service, as well as loans and other incentives. Some public utilities offer energy audits as part of a coordinated service to plan or install home energy upgrades. Utilities may also provide incentives to switch, for example, if you are an oil customer considering switching to natural gas.

Where to look for insulation recommendations:

- Local building inspector's office.

- Local or state building codes.

- US Department of Energy.

- Your local Builders Association.

Residential energy auditors are accredited by the Building Performance Institute (BPI) or the Residential Energy Services Network (RESNET).

There are also some simplified tools available, with which a homeowner can quickly assess energy improvement potential. Often these are supplied for free by state agencies or local utilities, who produce a report with estimates of usage by device/area (since they have usage information already). Examples include the Energy Trust of Oregon program and the Seattle Home Resource Profile. Such programs may also include free compact fluorescent lights.

A simple do-it-yourself home energy audit can be performed without using any specialized tools. With an attentive and planned assessment, a homeowner can spot many problems that cause energy losses and make decisions about possible energy efficiency upgrades. During a home energy audit it is important to have a checklist of areas that were inspected as well as problems identified. Once the audit is completed, a plan for suggested actions needs to be developed.

New York City

In New York City, local laws such as Local Law 87 require buildings larger than 50,000 square feet (4,600 m²) to have an energy audit once every ten years, as assigned by its parcel number. Energy auditors must be certified to perform this work, although there is no oversight to enforce the rule. Because Local Law 87 requires a licensed Professional Engineer to oversee the work, choosing a well-established engineering firm is the safest route.

These laws are the results of New York City's PlaNYC to reduce energy used by buildings, which is the greatest source of pollution in New York City. Some engineering firms provide free energy audits for facilities committed to implementing the energy saving measures found.

Lebanon

Since 2002, The Lebanese Center for Energy Conservation (LCEC) initiated a nationwide program on energy audits for medium and large consuming facilities. By the end of 2008, LCEC has financed and supervised more than 100 audits.

LCEC launched an energy audit program to assist Lebanese energy consuming tertiary and public buildings and industrial plants in the management of their energy through this program.

The long-term objective of LCEC is to create a market for ESCOs, whereby any beneficiary can contact directly a specialized ESCO to conduct an energy audit, implement energy conservation measures and monitor energy saving program according to a standardized energy performance contract.

Currently, LCEC is helping in the funding of the energy audit study and thus is linking both the beneficiary and the energy audit firm. LCEC also targets the creation of a special fund used for the implementation of the energy conservation measures resulting from the study.

LCEC set a minimum standard for the ESCOs qualifications in Lebanon.

Industrial Energy Audits

Increasingly in the last several decades, industrial energy audits have exploded as the demand to lower increasingly expensive energy costs and move towards a sustainable future have made energy audits greatly important. Their importance is magnified since energy spending is a major expense to industrial companies (energy spending accounts for ~ 10% of the average manufacturer's expenses). This growing trend should only continue as energy costs continue to rise.

While the overall concept is similar to a home or residential energy audit, industrial energy audits require a different skillset. Weatherproofing and insulating a house are the main focus of residential energy audits. For industrial applications, it is the HVAC, lighting, and production equipment that use the most energy, and hence are the primary focus of energy audits.

Types of Energy Audit

The term energy audit is commonly used to describe a broad spectrum of energy studies ranging from a quick walk-through of a facility to identify major problem areas to a comprehensive analysis of the implications of alternative energy efficiency measures sufficient to satisfy the financial criteria of sophisticated investors. Numerous audit procedures have been developed for non-residential (tertiary) buildings. Audit is required to identify the most efficient and cost-effective Energy Conservation Opportunities (ECOs) or Measures (ECMs). Energy conservation opportunities (or measures) can consist in more efficient use or of partial or global replacement of the existing installation.

When looking to the existing audit methodologies developed in IEA EBC Annex 11, by ASHRAE and by Krarti, it appears that the main issues of an audit process are:

- The analysis of building and utility data, including study of the installed equipment and analysis of energy bills;

- The survey of the real operating conditions;

- The understanding of the building behaviour and of the interactions with weather, occupancy and operating schedules;

- The selection and the evaluation of energy conservation measures;

- The estimation of energy saving potential;

- The identification of customer concerns and needs.

Common types/levels of energy audits are distinguished below, although the actual tasks performed and level of effort may vary with the consultant providing services under these broad headings. The only way to ensure that a proposed audit will meet your specific needs is to spell out those requirements in a detailed scope of work. Taking the time to prepare a formal solicitation will also assure the building owner of receiving competitive and comparable proposals.

Generally, four levels of analysis can be outlined:

- Level 0 – Benchmarking: This first analysis consists in a preliminary Whole Building Energy Use (WBEU) analysis based on the analysis of the historic utility use and costs and the comparison of the performances of the buildings to those of similar buildings. This benchmarking of the studied installation allows determining if further analysis is required;

- Level I – Walk-through audit: Preliminary analysis made to assess building energy efficiency to identify not only simple and low-cost improvements but also a list of energy conservation measures (ECMs, or energy conservation opportunities, ECOs) to orient the future detailed audit. This inspection is based on visual verifications, study of installed equipment and operating data and detailed analysis of recorded energy consumption collected during the benchmarking phase;

- Level II – Detailed/General energy audit: Based on the results of the pre-audit, this type of energy audit consists in energy use survey in order to provide a comprehensive analysis of the studied installation, a more detailed analysis of the facility, a breakdown of the energy use and a first quantitative evaluation of the ECOs/ECMs selected to correct the defects or improve the existing installation. This level of analysis can involve advanced on-site measurements and sophisticated computer-based simulation tools to evaluate precisely the selected energy retrofits;

- Level III – Investment-Grade audit: Detailed Analysis of Capital-Intensive Modifications focusing on potential costly ECOs requiring rigorous engineering study.

Benchmarking

The impossibility of describing all possible situations that might be encountered during an audit means that it is necessary to find a way of describing what constitutes good, average and bad energy performance across a range of situations. The aim of benchmarking is to answer this question. Benchmarking mainly consists in comparing the measured consumption with reference consumption of other similar buildings or generated by simulation tools to identify excessive or unacceptable running costs. As mentioned before, benchmarking is also necessary to identify buildings presenting interesting energy saving potential. An important issue in benchmarking is the use of performance indexes to characterize the building.

These indexes can be:

- Comfort indexes, comparing the actual comfort conditions to the comfort requirements;

- Energy indexes, consisting in energy demands divided by heated/conditioned area, allowing comparison with reference values of the indexes coming from regulation or similar buildings;

- Energy demands, directly compared to "reference" energy demands generated by means of simulation tools.

Typically, benchmarks are established based on the energy outlets (loads) within the building and are then further parsed into "base loads" and "weather sensitive loads". These are established through a simple regression analysis of energy consumption and demand (if metered) correlated to weather (temperature and degree - day) data during the period for which utility data is available. Aggregate base loads will represent as the intercept of this regression and the slope will typically represent the combination of building envelope conduction and infiltration losses less losses or gains from the base loads themselves. For example, while lighting is typically a base load, the heat generated from that lighting must be subtracted from the weather sensitive cooling load derived from the slope to gain an accurate picture of the true contribution of the building envelope on cooling energy use and demand.

Walk-through or Preliminary Audit

The preliminary audit (alternatively called a simple audit, screening audit or walk-through audit) is the simplest and quickest type of audit. It involves minimal interviews with site-operating personnel, a brief review of facility utility bills and other operating data, and a walk-through of the facility to become familiar with the building operation and to identify any glaring areas of energy waste or inefficiency.

Typically, only major problem areas will be covered during this type of audit. Corrective measures are briefly described, and quick estimates of implementation cost, potential operating cost savings, and simple payback periods are provided. A list of energy conservation measures (ECMs, or energy conservation opportunities, ECOs) requiring further consideration is also provided. This level of detail, while not sufficient for reaching a final decision on implementing proposed measure, is adequate to prioritize energy-efficiency projects and to determine the need for a more detailed audit.

General Audit

The general audit (alternatively called a mini-audit, site energy audit or detailed energy audit or complete site energy audit) expands on the preliminary audit described above by collecting more detailed information about facility operation and by performing a more detailed evaluation of energy conservation measures. Utility bills are collected for a 12- to 36-month period to allow the auditor to evaluate the facility's energy demand rate structures and energy usage profiles. If interval meter data is available, the detailed energy profiles that such data makes possible will typically be analyzed for signs of energy waste. Additional metering of specific energy-consuming systems is often performed to supplement utility data. In-depth interviews with facility operating personnel are conducted to provide a better understanding of major energy consuming systems and to gain insight into short- and longer-term energy consumption patterns. This type of audit will be able to identify all energy-conservation measures appropriate for the facility, given its operating parameters. A detailed financial analysis is performed for each measure based on detailed implementation cost estimates, site-specific operating cost savings, and the customer's investment criteria. Sufficient detail is provided to justify project implementation. The evolution of cloud-based energy auditing software platforms is enabling the managers of commercial buildings to collaborate with general and specialty trades contractors in performing general and energy system-specific audits. The benefit of software-enabled collaboration is the ability to identify the full range of energy efficiency options that may be applicable to the specific building under study with "live time" cost and benefit estimates supplied by local contractors.

Investment-grade Audit

In most corporate settings, upgrades to a facility's energy infrastructure must compete for capital funding with non-energy-related investments. Both energy and non-energy investments are rated on a single set of financial criteria that generally stress the expected return on investment (ROI). The projected operating savings from the implementation of energy projects must be developed such that they provide a high level of confidence. In fact, investors often demand guaranteed savings. The investment-grade audit expands on the detailed audit described above and relies on a complete engineering study in order to detail technical and economical issues necessary to justify the investment related to the transformations.

Simulation-based Energy Audit Procedure for Non-residential Buildings

A complete audit procedure, very similar to the ones proposed by ASHRAE and Krarti, has been proposed in the frame of the AUDITAC and HARMONAC projects to help in the implementation of the EPB ("Energy Performance of Buildings") directive in Europe and to fit to the current European market.

The following procedure proposes to make an intensive use of modern BES tools at each step of the audit process, from benchmarking to detailed audit and financial study:

- Benchmarking stage: While normalization is required to allow comparison between data recorded on the studied installation and reference values deduced from case studies or statistics. The use of simulation models, to perform a code-compliant simulation of the installation under study, allows to assess directly the studied installation, without any normalization needed. Indeed, applying a simulation-based benchmarking tool allows an individual normalization and allows avoiding size and climate normalization.

- Preliminary audit stage: Global monthly consumptions are generally insufficient to allow an accurate understanding of the building's behaviour. Even if the analysis of the energy bills does not allow identifying with accuracy the different energy consumers present in the facility, the consumption records can be used to calibrate building and system simulation models. To assess the existing system and to simulate correctly the building's thermal behaviour, the simulation model has to be calibrated on the studied installation. The iterations needed to perform the calibration of the model can also be fully integrated in the audit process and help in identifying required measurements and critical issues.

- Detailed audit stage: At this stage, on-site measurements, sub-metering and monitoring data are used to refine the calibration of the BES tool. Extensive attention is given to understanding not only the operating characteristics of all energy consuming systems, but also situations that cause load profile variations on short and longer term bases (e.g. daily, weekly, monthly, annual). When the calibration criteria is satisfied, the savings related to the selected ECOs/ECMs can be quantified.

- Investment-grade audit stage: At this stage, the results provided by the calibrated BES tool can be used to assess the selected ECOs/ECMs and orient the detailed engineering study.

Specific Audit Techniques

Infrared Thermography Audit

The advent of high-resolution thermography has enabled inspectors to identify potential issues within the building envelope by taking a thermal image of the various surfaces of a building. For purposes of an energy audit, the thermographer will analyze the patterns within the surface temperatures to identify heat transfer through convection, radiation, or conduction. It is important to note that the thermography *only* identifies *surface* temperatures, and analysis must be applied to determine the reasons for the patterns within the surface temperatures. Thermal analysis of a home generally costs between 300 and 600 dollars.

For those who cannot afford a thermal inspection, it is possible to get a general feel for the heat loss with a non-contact infrared thermometer and several sheets of reflective insulation. The method involves measuring the temperatures on the inside surfaces of several exterior walls to establish baseline temperatures. After this, reflective barrier insulation is taped securely to the walls in 8-foot (2.4 m) by 1.5-foot (0.46 m) strips and the temperatures are measured in the center of the insulated areas at 1-hour intervals for 12 hours (the reflective barrier is pulled away from the wall to measure the temperature in the center of the area which it has covered). The best manner in which to do this is when the temperature differential (Delta T) between the inside and outside of the structure is at least 40 degrees. A well-insulated wall will commonly change approximately 1 degree per hour if the difference between external and internal temperatures is an average of 40 degrees. A poorly insulated wall can drop as much as 10 degrees in an hour.

Pollution Audits

With increases in carbon dioxide emissions or other greenhouse gases, pollution audits are now a prominent factor in most energy audits. Implementing energy efficient technologies help prevent utility generated pollution.

Online pollution and emission calculators can help approximate the emissions of other prominent air pollutants in addition to carbon dioxide.

Pollution audits generally take electricity and heating fuel consumption numbers over a two-year period and provide approximations for carbon dioxide, VOCs, nitrous oxides, carbon monoxide, sulfur dioxide, mercury, cadmium, lead, mercury compounds, cadmium compounds and lead compounds.

Building Energy Rating Systems

- Australia – House Energy Rating.

- Canada – EnerGuide.

- UK – National Home Energy Rating, Standard Assessment Procedure, Energy Performance Certificate.

- US – Home energy rating, Energy Star.

Energy Harvesting

Energy harvesting (also known as power harvesting or energy scavenging or ambient power) is the process by which energy is derived from external sources (e.g., solar power, thermal energy, wind energy, salinity gradients, and kinetic energy, also known as ambient energy), captured, and stored for small, wireless autonomous devices, like those used in wearable electronics and wireless sensor networks.

Energy harvesters provide a very small amount of power for low-energy electronics. While the input fuel to some large-scale generation costs resources (oil, coal, etc.), the energy source for energy harvesters is present as ambient background. For example, temperature gradients exist from the operation of a combustion engine and in urban areas, there is a large amount of electromagnetic energy in the environment because of radio and television broadcasting.

One of the earliest applications of ambient power collected from ambient electromagnetic radiation (EMR) is the crystal radio.

The principles of energy harvesting from ambient EMR can be demonstrated with basic components.

Operation

Energy harvesting devices converting ambient energy into electrical energy have attracted much interest in both the military and commercial sectors. Some systems convert motion, such as that of ocean waves, into electricity to be used by oceanographic monitoring sensors for autonomous operation. Future applications may include high power output devices (or arrays of such devices) deployed at remote locations to serve as reliable power stations for large systems. Another application is in wearable electronics, where energy harvesting devices can power or recharge cellphones, mobile computers, radio communication equipment, etc. All of these devices must be sufficiently robust to endure long-term exposure to hostile environments and have a broad range of dynamic sensitivity to exploit the entire spectrum of wave motions.

Accumulating Energy

Energy can also be harvested to power small autonomous sensors such as those developed using MEMS technology. These systems are often very small and require little power, but their applications are limited by the reliance on battery power. Scavenging energy from ambient vibrations, wind, heat or light could enable smart sensors to be functional indefinitely.

Typical power densities available from energy harvesting devices are highly dependent upon the specific application (affecting the generator's size) and the design itself of the harvesting generator. In general, for motion powered devices, typical values are a few $\mu W/cm^3$ for human body powered applications and hundreds of $\mu W/cm^3$ for generators powered from machinery. Most energy scavenging devices for wearable electronics generate very little power.

Storage of Power

In general, energy can be stored in a capacitor, super capacitor, or battery. Capacitors are used when the application needs to provide huge energy spikes. Batteries leak less energy and are therefore used when the device needs to provide a steady flow of energy. Compared to batteries, super capacitors have virtually unlimited charge-discharge cycles and can therefore operate forever enabling a maintenance-free operation in IoT and wireless sensor devices.

Use of the Power

Current interest in low power energy harvesting is for independent sensor networks. In these applications an energy harvesting scheme puts power stored into a capacitor then boosted/regulated to a second storage capacitor or battery for the use in the microprocessor or in the data transmission. The power is usually used in a sensor application and the data stored or is transmitted possibly through a wireless method.

Motivation

The history of energy harvesting dates back to the windmill and the waterwheel. People have searched for ways to store the energy from heat and vibrations for many decades. One driving force behind the search for new energy harvesting devices is the desire to power sensor networks and mobile devices without batteries. Energy harvesting is also motivated by a desire to address the issue of climate change and global warming.

Devices

There are many small-scale energy sources that generally cannot be scaled up to industrial size:

- Some wristwatches are powered by kinetic energy (called automatic watches), in this case movement of the arm is used. The arm movement causes winding of its mainspring. A newer design introduced by Seiko ("Kinetic") uses movement of a magnet in the electromagnetic generator instead to power the quartz movement. The motion provides a rate of change of flux, which results in some induced emf on the coils. The concept is related to Faraday's Law.

- Photovoltaics is a method of generating electrical power by converting solar radiation (both indoors and outdoors) into direct current electricity using semiconductors that exhibit the photovoltaic effect. Photovoltaic power generation employs solar panels composed of a number of cells containing a photovoltaic material. Note that photovoltaics have been scaled up to industrial size and that large solar farms exist.

- Thermoelectric generators (TEGs) consist of the junction of two dissimilar materials and the presence of a thermal gradient. Large voltage outputs are possible by connecting many junctions electrically in series and thermally in parallel. Typical performance is 100-300 μV/K per junction. These can be utilized to capture mW.s of energy from industrial equipment, structures, and even the human body. They are typically coupled with heat sinks to improve temperature gradient.

- Micro wind turbine are used to harvest wind energy readily available in the environment in the form of kinetic energy to power the low power electronic devices such as wireless sensor nodes. When air flows across the blades of the turbine, a net pressure difference is developed between the wind speeds above and below the blades. This will result in a lift force generated which in turn rotate the blades. Similar to photovoltaics, wind farms have been constructed on an industrial scale and are being used to generate substantial amounts of electrical energy.

- Piezoelectric crystals or fibers generate a small voltage whenever they are mechanically deformed. Vibration from engines can stimulate piezoelectric materials, as can the heel of a shoe, or the pushing of a button.

- Special antennas can collect energy from stray radio waves, this can also be done with a Rectenna and theoretically at even higher frequency EM radiation with a Nantenna.

- Power from keys pressed during use of a portable electronic device or remote controller, using magnet and coil or piezoelectric energy converters, may be used to help power the device.

Ambient-radiation Sources

A possible source of energy comes from ubiquitous radio transmitters. Historically, either a large collection area or close proximity to the radiating wireless energy source is needed to get useful power levels from this source. The nantenna is one proposed development which would overcome this limitation by making use of the abundant natural radiation (such as solar radiation).

One idea is to deliberately broadcast RF energy to power and collect information from remote devices: This is now commonplace in passive radio-frequency identification (RFID) systems, but the Safety and US Federal Communications Commission (and equivalent bodies worldwide) limit the maximum power that can be transmitted this way to civilian use. This method has been used to power individual nodes in a wireless sensor network.

Fluid Flow

Airflow can be harvested by various turbine and non-turbine generator technologies. For example, Zephyr Energy Corporation's patented Windbeam micro generator captures energy from airflow to recharge batteries and power electronic devices. The Windbeam's novel design allows it to operate silently in wind speeds as low as 2 mph. The generator consists of a lightweight beam suspended by durable long-lasting springs within an outer frame. The beam oscillates rapidly when exposed to airflow due to the effects of multiple fluid flow phenomena. A linear alternator assembly converts the oscillating beam motion into usable electrical energy. A lack of bearings and gears eliminates frictional inefficiencies and noise. The generator can operate in low-light environments unsuitable for solar panels (e.g. HVAC ducts) and is inexpensive due to low cost components and simple construction. The scalable technology can be optimized to satisfy the energy requirements and design constraints of a given application.

The flow of blood can also be used to power devices. For instance, the pacemaker developed at the University of Bern, uses blood flow to wind up a spring which in turn drives an electrical micro-generator.

Photovoltaic

Photovoltaic (PV) energy harvesting wireless technology offers significant advantages over wired or solely battery-powered sensor solutions: virtually inexhaustible sources of power with little or no adverse environmental effects. Indoor PV harvesting solutions have to date been powered by specially tuned amorphous silicon (aSi)a technology most used in Solar Calculators. In recent years new PV technologies have come to the forefront in Energy Harvesting such as Dye Sensitized Solar Cells (DSSC). The dyes absorbs light much like chlorophyll does in plants. Electrons released on impact escape to the layer of TiO_2 and from there diffuse, through the electrolyte, as the dye can be tuned to the visible spectrum much higher power can be produced. At 200 lux a DSSC can provide over 10 µW per cm².

Piezoelectric

The piezoelectric effect converts mechanical strain into electric current or voltage. This strain can come from many different sources. Human motion, low-frequency seismic vibrations, and acoustic noise are everyday examples. Except in rare instances the piezoelectric effect operates in AC requiring time-varying inputs at mechanical resonance to be efficient.

Batteryless and wireless wallswitch.

Most piezoelectric electricity sources produce power on the order of milliwatts, too small for system application, but enough for hand-held devices such as some commercially available self-winding wristwatches. One proposal is that they are used for micro-scale devices, such as in a device harvesting micro-hydraulic energy. In this device, the flow of pressurized hydraulic fluid drives a reciprocating piston supported by three piezoelectric elements which convert the pressure fluctuations into an alternating current.

As piezo energy harvesting has been investigated only since the late 1990s, it remains an emerging technology. Nevertheless, some interesting improvements were made with the self-powered electronic switch at INSA school of engineering, implemented by the spin-off Arveni. In 2006, the proof of concept of a battery-less wireless doorbell push button was created, and recently, a product showed that classical wireless wallswitch can be powered by a piezo harvester. Other industrial applications appeared between 2000 and 2005, to harvest energy from vibration and supply sensors for example, or to harvest energy from shock.

Piezoelectric systems can convert motion from the human body into electrical power. DARPA has funded efforts to harness energy from leg and arm motion, shoe impacts, and blood pressure for low level power to implantable or wearable sensors. The nanobrushes are another example of a piezoelectric energy harvester. They can be integrated into clothing. Multiple other nanostructures have been exploited to build an energy-harvesting device, for example, a single crystal PMN-PT nanobelt was fabricated and assembled into a piezoelectric energy harvester in 2016. Careful design is needed to minimise user discomfort. These energy harvesting sources by association affect the body. The Vibration Energy Scavenging Project is another project that is set up to try to scavenge electrical energy from environmental vibrations and movements. Microbelt can be used to gather electricity from respiration. Besides, as the vibration of motion from human comes in three directions, a single piezoelectric cantilever based omni-directional energy harvester is created by using 1:2 internal resonance. Finally, a millimeter-scale piezoelectric energy harvester has also already been created.

The use of piezoelectric materials to harvest power has already become popular. Piezoelectric materials have the ability to transform mechanical strain energy into electrical charge. Piezo elements are being embedded in walkways to recover the "people energy" of footsteps. They can also be embedded in shoes to recover "walking energy". Researchers at MIT developed the first micro-scale piezoelectric energy harvester using thin film PZT in 2005. Arman Hajati and Sang-Gook Kim invented the Ultra Wide-Bandwidth micro-scale piezoelectric energy harvesting device by exploiting the nonlinear stiffness of a doubly clamped microelectromechanical systems (MEMSs) resonator. The stretching strain in a doubly clamped beam shows a nonlinear stiffness, which provides a passive feedback and results in amplitude-stiffened Duffing mode resonance. Typically, piezoelectric cantilevers are adopted for the above-mentioned energy harvesting system. One drawback is that the piezoelectric cantilever has gradient strain distribution, i.e., the piezoelectric transducer is not fully utilized. To address this issue, triangle shaped and L-shaped cantilever are proposed for uniform strain distribution.

In 2018, Soochow University researchers reported hybridizing a triboelectric nanogenerator and a silicon solar cell by sharing a mutual electrode. This device can collect solar energy *or* convert the mechanical energy of falling raindrops into electricity.

Energy from Smart Roads and Piezoelectricity

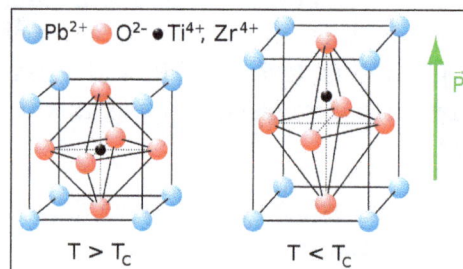

Tetragonal unit cell of lead titanate.

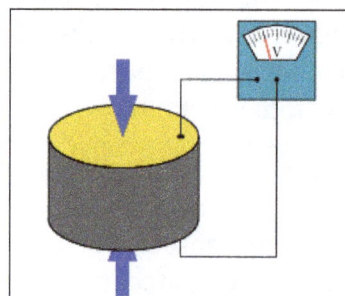

A piezoelectric disk generates a voltage when deformed (change in shape is greatly exaggerated).

Brothers Pierre Curie and Jacques Curie gave the concept of piezoelectric effect in 1880. Piezoelectric effect converts mechanical strain into voltage or electric current and generates electric energy from motion, weight, vibration and temperature changes as shown in the figure.

Considering piezoelectric effect in thin film lead zirconate titanate PZT, microelectromechanical systems (MEMS) power generating device has been developed. During recent improvement in piezoelectric technology, Aqsa Abbasi (also known as Aqsa Aitbar, General secretory at IMS, IEEE MUET Chapter and Director Media at HYD MUN) diffentiated two modes called and in vibration converters and re-designed to resonate at specific frequencies from an external vibration energy source, thereby creating electrical energy via the piezoelectric effect using electromechanical damped mass. However, Aqsa further developed beam-structured electrostatic devices that are more difficult to fabricate than PZT MEMS devices versus a similar because general silicon processing involves many more mask steps that do not require PZT film. Piezoelectric type sensors and actuators have a cantilever beam structure that consists of a membrane bottom electrode, film, piezoelectric film, and top electrode. More than (3~5 masks) mask steps are required for patterning of each layer while have very low induced voltage. Pyroelectric crystals that have a unique polar axis and have spontaneous polarization, along which the spontaneous polarization exists. These are the crystals of classes 6mm, 4mm, mm2, 6, 4, 3m, 3,2, m. The special polar axis—crystallophysical axis $X3$ — coincides with the axes $L6$, $L4$, $L3$, and $L2$ of the crystals or lies in the unique straight plane P (class "m"). Consequently, the electric centers of positive and negative charges are displaced of an elementary cell from equilibrium positions, i.e., the spontaneous polarization of the crystal changes. Therefore, all considered crystals have spontaneous polarization Since piezoelectric effect in pyroelectric crystals arises as a result of changes in their spontaneous polarization under external effects (electric fields, mechanical stresses). As a result of displacement, Aqsa Abbasi introduced change in the components along all three axes Suppose that is proportional to the mechanical stresses causing in a first approximation, which results where Tkl represents the mechanical stress and $dikl$ represents the piezoelectric modules.

PZT thin films have attracted attention for applications such as force sensors, accelerometers, gyroscopes actuators, tunable optics, micro pumps, ferroelectric RAM, display systems and smart roads, when energy sources are limited, energy harvesting plays an important role in the environment. Smart roads have the potential to play an important role in power generation. Embedding piezoelectric material in the road can convert pressure exerted by moving vehicles into voltage and current.

Smart Transportation Intelligent System

Piezoelectric sensors are most useful in Smart-road technologies that can be used to create systems that are intelligent and improve productivity in the long run. Imagine highways that alert motorists of a traffic jam before it forms. Or bridges that report when they are at risk of collapse, or an electric grid that fixes itself when blackouts hit. For many decades, scientists and experts have argued that the best way to fight congestion is intelligent transportation systems, such as roadside sensors to measure traffic and synchronized traffic lights to control the flow of vehicles. But the spread of these technologies has been limited by cost. There are also some other smart-technology shovel ready projects which could be deployed fairly quickly, but most of the technologies are still at the development stage and might not be practically available for five years or more.

Pyroelectric

The pyroelectric effect converts a temperature change into electric current or voltage. It is analogous to the piezoelectric effect, which is another type of ferroelectric behavior. Pyroelectricity requires time-varying inputs and suffers from small power outputs in energy harvesting applications due to its low operating frequencies. However, one key advantage of pyroelectrics over thermoelectrics is that many pyroelectric materials are stable up to 1200 °C or higher, enabling energy harvesting from high temperature sources and thus increasing thermodynamic efficiency.

One way to directly convert waste heat into electricity is by executing the Olsen cycle on pyroelectric materials. The Olsen cycle consists of two isothermal and two isoelectric field processes in the electric displacement-electric field (D-E) diagram. The principle of the Olsen cycle is to charge a capacitor via cooling under low electric field and to discharge it under heating at higher electric field. Several pyroelectric converters have been developed to implement the Olsen cycle using conduction, convection, or radiation. It has also been established theoretically that pyroelectric conversion based on heat regeneration using an oscillating working fluid and the Olsen cycle can reach Carnot efficiency between a hot and a cold thermal reservoir. Moreover, recent studies have established polyvinylidene fluoride trifluoroethylene [P(VDF-TrFE)] polymers and lead lanthanum zirconate titanate (PLZT) ceramics as promising pyroelectric materials to use in energy converters due to their large energy densities generated at low temperatures. Additionally, a pyroelectric scavenging device that does not require time-varying inputs was recently introduced. The energy-harvesting device uses the edge-depolarizing electric field of a heated pyroelectric to convert heat energy into mechanical energy instead of drawing electric current off two plates attached to the crystal-faces.

Thermoelectrics

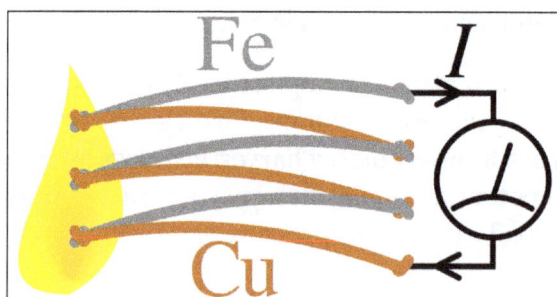

Seebeck effect in a thermopile made from iron and copper wires.

In 1821, Thomas Johann Seebeck discovered that a thermal gradient formed between two dissimilar conductors produces a voltage. At the heart of the thermoelectric effect is the fact that a temperature gradient in a conducting material results in heat flow; this results in the diffusion of charge carriers. The flow of charge carriers between the hot and cold regions in turn creates a voltage difference. In 1834, Jean Charles Athanase Peltier discovered that running an electric current through the junction of two dissimilar conductors could, depending on the direction of the current, cause it to act as a heater or cooler. The heat absorbed or produced is proportional to the current, and the proportionality constant is known as the Peltier coefficient. Today, due to knowledge of the Seebeck and Peltier effects, thermoelectric materials can be used as heaters, coolers and generators (TEGs).

Ideal thermoelectric materials have a high Seebeck coefficient, high electrical conductivity, and low thermal conductivity. Low thermal conductivity is necessary to maintain a high thermal gradient at the junction. Standard thermoelectric modules manufactured today consist of P- and N-doped bismuth-telluride semiconductors sandwiched between two metallized ceramic plates. The ceramic plates add rigidity and electrical insulation to the system. The semiconductors are connected electrically in series and thermally in parallel.

Miniature thermocouples have been developed that convert body heat into electricity and generate 40 µW at 3 V with a 5 degree temperature gradient, while on the other end of the scale, large thermocouples are used in nuclear RTG batteries.

Practical examples are the finger-heartratemeter by the Holst Centre and the thermogenerators by the Fraunhofer Gesellschaft.

Advantages to thermoelectrics:

1.   No moving parts allow continuous operation for many years. *Tellurex Corporation* (a thermoelectric production company) claims that thermoelectrics are capable of over 100,000 hours of steady state operation.

2.   Thermoelectrics contain no materials that must be replenished.

3.   Heating and cooling can be reversed.

One downside to thermoelectric energy conversion is low efficiency (currently less than 10%). The development of materials that are able to operate in higher temperature gradients, and that can conduct electricity well without also conducting heat (something that was until recently thought impossible), will result in increased efficiency.

Future work in thermoelectrics could be to convert wasted heat, such as in automobile engine combustion, into electricity.

Electrostatic (Capacitive)

This type of harvesting is based on the changing capacitance of vibration-dependent capacitors. Vibrations separate the plates of a charged variable capacitor, and mechanical energy is converted into electrical energy. Electrostatic energy harvesters need a polarization source to work and to convert mechanical energy from vibrations into electricity. The polarization source should be in the order of some hundreds of volts; this greatly complicates the power management circuit. Another solution consists in using electrets, that are electrically charged dielectrics able to keep the polarization on the capacitor for years. It's possible to adapt structures from classical electrostatic induction generators, which also extract energy from variable capacitances, for this purpose. The resulting devices are self-biasing, and can directly charge batteries, or can produce exponentially growing voltages on storage capacitors, from which energy can be periodically extracted by DC/DC converters.

Magnetic Induction

Magnets wobbling on a cantilever are sensitive to even small vibrations and generate microcurrents by moving relative to conductors due to Faraday's law of induction. By developing a miniature device of this kind in 2007, a team from the University of Southampton made possible the planting of such a device in environments that preclude having any electrical connection to the outside world. Sensors in inaccessible places can now generate their own power and transmit data to outside receivers.

One of the major limitations of the magnetic vibration energy harvester developed at University of Southampton is the size of the generator, in this case approximately one cubic centimeter, which is much too large to integrate into today's mobile technologies. The complete generator including circuitry is a massive 4 cm by 4 cm by 1 cm nearly the same size as some mobile devices such as the iPod nano. Further reductions in the dimensions are possible through the integration of new and more flexible materials as the cantilever beam component. In 2012, a group at Northwestern University developed a vibration-powered generator out of polymer in the form of a spring. This device was able to target the same frequencies as the University of Southampton groups silicon based device but with one third the size of the beam component.

A new approach to magnetic induction based energy harvesting has also been proposed by using ferrofluids. Quite recently, the change in domain wall pattern with the application of stress has been proposed as a method to harvest energy using magnetic induction. In this, the authors have shown that the applied stress can change the domain pattern in microwires. Ambient vibrations can cause stress in microwires, which can induce a change in domain pattern and hence change the induction. Power, of the order of uW/cm² has been reported.

Commercially successful vibration energy harvesters based on magnetic induction are still relatively few in number. Examples include products developed by Swedish company ReVibe Energy, a technology spin-out from Saab Group. Another example is the products developed from the early University of Southampton prototypes by Perpetuum. These have to be sufficiently large to generate the power required by wireless sensor nodes (wsn)but in M2M applications this is not normally an issue. These harvesters are now being supplied in large volumes to power wsn's made by companies such as GE and Emerson and also for train bearing monitoring systems made by Perpetuum. Overhead powerline sensors can use magnetic induction to harvest energy directly from the conductor they are monitoring.

Blood Sugar

Another way of energy harvesting is through the oxidation of blood sugars. These energy harvesters are called biobatteries. They could be used to power implanted electronic devices (e.g., pacemakers, implanted biosensors for diabetics, implanted active RFID devices, etc.) At present, the Minteer Group of Saint Louis University has created enzymes that could be used to generate power from blood sugars. However, the enzymes would still need to be replaced after a few years. In 2012, a pacemaker was powered by implantable biofuel cells at Clarkson University under the leadership of Dr. Evgeny Katz.

Tree-based

Tree metabolic energy harvesting is a type of bio-energy harvesting. Voltree has developed a method for harvesting energy from trees. These energy harvesters are being used to power remote sensors and mesh networks as the basis for a long term deployment system to monitor forest fires and weather in the forest. According to Voltree's website, the useful life of such a device should be limited only by the lifetime of the tree to which it is attached. A small test network was recently deployed in a US National Park forest.

Other sources of energy from trees include capturing the physical movement of the tree in a generator. Theoretical analysis of this source of energy shows some promise in powering small electronic devices. A practical device based on this theory has been built and successfully powered a sensor node for a year.

Metamaterial

A metamaterial-based device wirelessly converts a 900 MHz microwave signal to 7.3 volts of direct current (greater than that of a USB device). The device can be tuned to harvest other signals including Wi-Fi signals, satellite signals, or even sound signals. The experimental device used a series of five fiberglass and copper conductors. Conversion efficiency reached 37 percent. When traditional antennas are close to each other in space they interfere with each other. But since RF power goes down by the cube of the distance, the amount of power is very very small. While the claim of 7.3 volts is grand, the measurement is for an open circuit. Since the power is so low, there can be almost no current when any load is attached.

Atmospheric Pressure Changes

The pressure of the atmosphere changes naturally over time from temperature changes and weather patterns. Devices with a sealed chamber can use these pressure differences to extract energy. This has been used to provide power for mechanical clocks such as the Atmos clock.

Ocean Energy

A relatively new concept of generating energy is to generate energy from oceans. Large masses of waters are present on the planet which carry with them great amounts of energy. The energy in this case can be generated by tidal streams, ocean waves, difference in salinity and also difference in temperature. As of 2018, efforts are underway to harvest energy this way. United States Navy recently was able to generate electricity using difference in temperatures present in the ocean.

Future Directions

Electroactive polymers (EAPs) have been proposed for harvesting energy. These polymers have a large strain, elastic energy density, and high energy conversion efficiency. The total weight of systems based on EAPs (electroactive polymers) is proposed to be significantly lower than those based on piezoelectric materials.

Nanogenerators, such as the one made by Georgia Tech, could provide a new way for powering devices without batteries. As of 2008, it only generates some dozen nanowatts, which is too low for any practical application.

Noise has been the subject of a proposal by NiPS Laboratory in Italy to harvest wide spectrum low scale vibrations via a nonlinear dynamical mechanism that can improve harvester efficiency up to a factor 4 compared to traditional linear harvesters.

Combinations of different types of energy harvesters can further reduce dependence on batteries, particularly in environments where the available ambient energy types change periodically. This type of complementary balanced energy harvesting has the potential to increase reliability of wireless sensor systems for structural health monitoring.

Alternative Energy

The potential issues surrounding the use of fossil fuels, particularly in terms of climate change, were considered earlier than you may think. It was a Swedish scientist named Svante Arrhenius who was the first to state that the use of fossil fuel could contribute to global warming, way back in 1896.

The issue has become a hot-button topic over the course of the last few decades. Today, there is a general shift towards environmental awareness and the sources of our energy are coming under closer scrutiny.

This has led to the rise of a number of alternative energy sources. While the viability of each can be argued, they all contribute something positive when compared to fossil fuels.

Lower emissions, lower fuel prices and the reduction of pollution are all advantages that the use of alternative fuels can often provide.

Examples of Alternative Energy Sources

Hydrogen Gas

Unlike other forms of natural gas, hydrogen is a completely clean burning fuel. Once produced, hydrogen gas cells emit only water vapor and warm air when in use.

The major issue with this form of alternative energy is that it is mostly derived from the use of natural gas and fossil fuels. As such, it could be argued that the emissions created to extract it counteract the benefits of its use.

The process of electrolysis, which is essential for the splitting of water into hydrogen and oxygen, makes this less of an issue. However, electrolysis still ranks below the previously mentioned methods for obtaining hydrogen, though research continues to make it more efficient and cost-effective.

Tidal Energy

While tidal energy uses the power of water to generate energy, much like with hydroelectric methods, its application actually has more in common with wind turbines in many cases.

Though it is a fairly new technology, its potential is enormous. A report produced in the United Kingdom estimated that tidal energy could meet as much as 20% of the UK's current electricity demands.

The most common form of tidal energy generation is the use of Tidal Stream Generators. These use the kinetic energy of the ocean to power turbines, without producing the waste of fossil fuels or being as susceptible to the elements as other forms of alternative energy.

Biomass Energy

Biomass energy comes in a number of forms. Burning wood has been used for thousands of years to create heat, but more recent advancements have also seen waste, such as that in landfills, and alcohol products used for similar purposes.

Focusing on burning wood, the heat generated can be equivalent to that of a central heating system. Furthermore, the costs involved tend to be lower and the amount of carbon released by this kind of fuel falls below the amount released by fossil fuels.

However, there are a number of issues that you need to consider with these systems, especially if installed in the home. Maintenance can be a factor, plus you may need to acquire permission from a local authority to install one.

Wind Energy

This form of energy generation has become increasingly popular in recent years. It offers much the same benefits that many other alternative fuel sources do in that it makes use of a renewable source and generates no waste.

Current wind energy installations power roughly twenty million homes in the United States per year and that number is growing. Most states in the nation now have some form of wind energy set-up and investment into the technology continues to grow.

Unfortunately, this form of energy generation also presents challenges. Wind turbines restrict views and may be dangerous to some forms of wildlife.

Geothermal Power

At its most basic, geothermal power is about extracting energy from the ground around us. It is growing increasingly popular, with the sector as a whole experiencing five percent growth in 2015.

The World Bank currently estimates that around forty countries could meet most of their power demands using geothermal power.

This power source has massive potential while doing little to disrupt the land. However, the heavy upfront costs of creating geothermal power plants has led to slower adoption than may have been expected for a fuel source with so much promise.

Natural Gas

Natural gas sources have been in use for a number of decades, but it is through the progression of compression techniques that it is becoming a more viable alternative energy source. In particular, it is being used in cars to reduce carbon emissions.

Demand for this energy source has been increasing. In 2016, the lower 48 states of the United States reached record levels of demand and consumption.

Despite this, natural gas does come with some issues. The potential for contamination is larger than with other alternative fuel sources and natural gas still emits greenhouse gases, even if the amount is lower than with fossil fuels.

Biofuels

In contrast to biomass energy sources, biofuels make use of animal and plant life to create energy. In essence they are fuels that can be obtained from some form of organic matter.

They are renewable in cases where plants are used, as these can be regrown on a yearly basis. However, they do require dedicated machinery for extraction, which can contribute to increased emissions even if biofuels themselves don't.

Biofuels are increasingly being adopted, particularly in the United States. They accounted for approximately seven percent of transport fuel consumption as of 2012.

Wave Energy

Water again proves itself to be a valuable contributor to alternative energy fuel sources with wave energy converters. These hold an advantage over tidal energy sources because they can be placed in the ocean in various situations and locations.

Much like with tidal energy, the benefits come in the lack of waste produced. It is also more reliable than many other forms of alternative energy and has enormous potential when used properly.

Again, the cost of such systems is a major contributing factor to slow uptake. We also don't yet have enough data to find out how wave energy converters affect natural ecosystems.

Hydroelectric Energy

Hydroelectric methods actually are some of the earliest means of creating energy, though their use began to decline with the rise of fossil fuels. Despite this, they still account for approximately seven percent of the energy produced in the United States.

Hydroelectric energy carries with it a number of benefits. Not only is it a clean source of energy, which means it doesn't create pollution and the myriad issues that arise from it, but it is also a renewable energy source.

Better yet, it also offers a number of secondary benefits that are not immediately apparent. The dams used in generating hydroelectric power also contribute to flood control and irrigation techniques.

Nuclear Power

Nuclear power is amongst the most abundant forms of alternative energy. It creates a number of direct benefits in terms of emissions and efficiency, while also boosting the economy by creating jobs in plant creation and operation.

Thirteen countries relied on nuclear power to produce at least a quarter of their electricity as of 2015 and there are currently 450 plants in operation throughout the world.

The drawback is that when something goes wrong with a nuclear power plant the potential for catastrophe exists. The situations in Chernobyl and Fukushima are examples of this.

Solar Power

When most people think of alternative energy sources they tend to use solar power as an example. The technology has evolved massively over the years and is now used for large-scale energy production and power generation for single homes.

A number of countries have introduced initiatives to promote the growth of solar power. The United Kingdom's 'Feed-in Tariff' is one example, as is the United States' 'Solar Investment Tax Credit'.

This energy source is completely renewable and the costs of installation are outweighed by the money saved in energy bills from traditional suppliers. Nevertheless, solar cells are prone to deterioration over large periods of time and are not as effective in unideal weather conditions.

As the issues that result from the use of traditional fossil fuels become more prominent, alternative fuel sources like the ones mentioned here are likely to gain further importance.

Their benefits alleviate many of the problems caused by fossil fuel use, particularly when it comes to emissions. However, the advancement of some of these technologies has been slowed down due to the amount of investment needed to make them viable.

Through combining them all we may be able to positively affect issues like climate change, pollution and many others.

Green Building

Green Building, also known as green construction or sustainable building, is the practice of creating structures and using processes that are environmentally responsible and resource-efficient throughout a building's life-cycle: From siting to design, construction, operation, maintenance, renovation, and deconstruction. This practice expands and complements the classical building design concerns of economy, utility, durability, and comfort.

Although new technologies are constantly being developed to complement current practices in creating greener structures, the common objective is that green buildings are designed to reduce the overall impact of the built environment on human health and the natural environment by:

- Efficiently using energy, water, and other resources.

- Protecting occupant health and improving employee productivity.

- Reducing waste, pollution and environmental degradation.

A similar concept is natural building, which is usually on a smaller scale and tends to focus on the use of natural materials that are available locally. Other related topics include sustainable design and green architecture.

Reducing Environmental Impact

Green building practices aim to reduce the environmental impact of buildings. Buildings account for a large amount of land use, energy and water consumption, and air and atmosphere alteration. Considering the statistics, reducing the amount of natural resources buildings consume and the amount of pollution given off is seen as crucial for future sustainability, according to EPA. The environmental impact of buildings is often underestimated, while the perceived costs of green buildings are overestimated. A recent survey by the World Business Council for Sustainable Development finds that green costs are overestimated by 300 percent, as key players in real estate and construction estimate the additional cost at 17 percent above conventional construction, more than triple the true average cost difference of about 5 percent.

Goals of Green Building

The concept of sustainable development can be traced to the energy (especially fossil oil) crisis and the environment pollution concern in the 1970s. The green building movement in the U.S. originated from the need and desire for more energy efficient and environmentally friendly construction practices. There are a number of motives to building green, including environmental, economic, and social benefits. However, modern sustainability initiatives call for an integrated and synergistic design to both new construction and in the retrofitting of an existing structure. Also known as sustainable design, this approach integrates the building life-cycle with each green practice employed with a design-purpose to create a synergy amongst the practices used.

Green building brings together a vast array of practices and techniques to reduce and ultimately eliminate the impacts of buildings on the environment and human health. It often emphasizes taking advantage of renewable resources, e.g., using sunlight through passive solar, active solar, and photovoltaic techniques and using plants and trees through green roofs, rain gardens, and for reduction of rainwater run-off. Many other techniques, such as using packed gravel or permeable concrete instead of conventional concrete or asphalt to enhance replenishment of ground water, are used as well.

While the practices, or technologies, employed in green building are constantly evolving and may differ from region to region, there are fundamental principles that persist from which the method is derived: Siting and Structure Design Efficiency, Energy Efficiency, Water Efficiency, Materials Efficiency, Indoor Environmental Quality Enhancement, Operations and Maintenance Optimization, and Waste and Toxics Reduction. The essence of green building is an optimization of one or more of these principles. Also, with the proper synergistic design, individual green building technologies may work together to produce a greater cumulative effect.

On the aesthetic side of green architecture or sustainable design is the philosophy of designing a building that is in harmony with the natural features and resources surrounding the site. There are several key steps in designing sustainable buildings: specify 'green' building materials from local sources, reduce loads, optimize systems, and generate on-site renewable energy.

Siting and Structure Design Efficiency

The foundation of any construction project is rooted in the concept and design stages. The concept stage, in fact, is one of the major steps in a project life cycle, as it has the largest impact on cost and performance. In designing environmentally optimal buildings, the objective function aims at minimizing the total environmental impact associated with all life-cycle stages of the building project. However, building as a process is not as streamlined as an industrial process, and varies from one building to the other, never repeating itself identically. In addition, buildings are much more complex products, composed of a multitude of materials and components each constituting various design variables to be decided at the design stage. A variation of every design variable may affect the environment during all the building's relevant life-cycle stages.

Energy Efficiency

Green buildings often include measures to reduce energy use. To increase the efficiency of the building envelope, (the barrier between conditioned and unconditioned space), they may use high-efficiency windows and insulation in walls, ceilings, and floors. Another strategy, passive solar building design, is often implemented in low-energy homes. Designers orient windows and walls and place awnings, porches, and trees to shade windows and roofs during the summer while maximizing solar gain in the winter. In addition, effective window placement (daylighting) can provide more natural light and lessen the need for electric lighting during the day. Solar water heating further reduces energy loads.

Onsite generation of renewable energy through solar power, wind power, hydro power, or biomass can significantly reduce the environmental impact of the building. Power generation is generally the most expensive feature to add to a building.

Water Efficiency

Reducing water consumption and protecting water quality are key objectives in sustainable building. One critical issue of water consumption is that in many areas of the country, the demands on the supplying aquifer exceed its ability to replenish itself. To the maximum extent feasible, facilities should increase their dependence on water that is collected, used, purified, and reused on-site. The protection and conservation of water throughout the life of a building may be accomplished by designing for dual plumbing that recycles water in toilet flushing. Waste-water may be minimized by utilizing water conserving fixtures such as ultra-low flush toilets and low-flow shower heads. Bidets help eliminate the use of toilet paper, reducing sewer traffic and increasing possibilities of re-using water on-site. Point of use water treatment and heating improves both water quality and energy efficiency while reducing the amount of water in circulation. The use of non-sewage and greywater for on-site use such as site-irrigation will minimize demands on the local aquifer.

Materials Efficiency

Building materials typically considered to be 'green' include rapidly renewable plant materials like bamboo (because bamboo grows quickly) and straw, lumber from forests certified to be sustainably managed, ecology blocks, dimension stone, recycled stone, recycled metal, and other products that are non-toxic, reusable, renewable, and/or recyclable (e.g. Trass, Linoleum, sheep wool, panels made from paper flakes, compressed earth block, adobe, baked earth, rammed earth, clay, vermiculite, flax linen, sisal, seagrass, cork, expanded clay grains, coconut, wood fibre plates, calcium sand stone, concrete (high and ultra high performance, roman self-healing concrete), etc.) The EPA (Environmental Protection Agency) also suggests using recycled industrial goods, such as coal combustion products, foundry sand, and demolition debris in construction projects Polyurethane heavily reduces carbon emissions as well. Polyurethane blocks are being used instead of CMTs by companies like American Insulock. Polyurethane blocks provide more speed, less cost, and they are environmentally friendly. Building materials should be extracted and manufactured locally to the building site to minimize the energy embedded in their transportation. Where possible, building elements should be manufactured off-site and delivered to site, to maximise benefits of off-site manufacture including minimising waste, maximising recycling (because manufacture is in one location), high quality elements, better OHS management, less noise and dust.

Indoor Environmental Quality Enhancement

The Indoor Environmental Quality (IEQ) category in LEED standards, one of the five environmental categories, was created to provide comfort, well-being, and productivity of occupants. The LEED IEQ category addresses design and construction guidelines especially: indoor air quality (IAQ), thermal quality, and lighting quality.

Indoor Air Quality seeks to reduce volatile organic compounds, or VOC's, and other air impurities such as microbial contaminants. Buildings rely on a properly designed HVAC system to provide adequate ventilation and air filtration as well as isolate operations (kitchens, dry cleaners, etc.) from other occupancies. During the design and construction process choosing construction materials and interior finish products with zero or low emissions will improve IAQ. Many building materials and cleaning/maintenance products emit toxic gases, such as VOC's and formaldehyde. These gases can have a detrimental impact on occupants' health and productivity as well. Avoiding these products will increase a building's IEQ.

Personal temperature and airflow control over the HVAC system coupled with a properly designed building envelope will also aid in increasing a building's thermal quality. Creating a high performance luminous environment through the careful integration of natural and artificial light sources will improve on the lighting quality of a structure.

Operations and Maintenance Optimization

No matter how sustainable a building may have been in its design and construction, it can only remain so if it is operated responsibly and maintained properly. Ensuring operations and maintenance(O&M) personnel are part of the project's planning and development process will help retain the green criteria designed at the onset of the project. Every aspect of green building is integrated into the O&M phase of a building's life. The addition of new green technologies also falls on the O&M staff. Although the goal of waste reduction may be applied during the design, construction and demolition phases of a building's life-cycle, it is in the O&M phase that green practices such as recycling and air quality enhancement take place.

Waste Reduction

Green architecture also seeks to reduce waste of energy, water and materials used during construction. For example, in California nearly 60% of the state's waste comes from commercial buildings During the construction phase, one goal should be to reduce the amount of material going to landfills. Well-designed buildings also help reduce the amount of waste generated by the occupants as well, by providing on-site solutions such as compost bins to reduce matter going to landfills.

To reduce the impact on wells or water treatment plants, several options exist. "Greywater", wastewater from sources such as dishwashing or washing machines, can be used for subsurface irrigation, or if treated, for non-potable purposes, e.g., to flush toilets and wash cars. Rainwater collectors are used for similar purposes.

Centralized wastewater treatment systems can be costly and use a lot of energy. An alternative to this process is converting waste and wastewater into fertilizer, which avoids these costs and shows other benefits. By collecting human waste at the source and running it to a semi-centralized biogas plant with other biological waste, liquid fertilizer can be produced. This concept was demonstrated by a settlement in Lubeck Germany in the late 1990s. Practices like these provide soil with organic nutrients and create carbon sinks that remove carbon dioxide from the atmosphere, offsetting greenhouse gas emission. Producing artificial fertilizer is also more costly in energy than this process.

Cost

The most criticized issue about constructing environmentally friendly buildings is the price. Photo-voltaics, new appliances, and modern technologies tend to cost more money. Most green buildings cost a premium of <2%, but yield 10 times as much over the entire life of the building. The stigma is between the knowledge of up-front cost vs. life-cycle cost. The savings in money come from more efficient use of utilities which result in decreased energy bills. Also, higher worker or student productivity can be factored into savings and cost deductions. Studies have shown over a 20 year life period, some green buildings have yielded $53 to $71 per square foot back on investment. It is projected that different sectors could save $130 Billion on energy bills.

Green Computing

Green computing, green ICT as per International Federation of Global & Green ICT "IFGICT", green IT, or ICT sustainability, is the study and practice of environmentally sustainable computing or IT.

The goals of green computing are similar to green chemistry: reduce the use of hazardous materials, maximize energy efficiency during the product's lifetime, the recyclability or biodegradability of defunct products and factory waste. Green computing is important for all classes of systems, ranging from handheld systems to large-scale data centers.

Many corporate IT departments have green computing initiatives to reduce the environmental effect of their IT operations.

In 1992, the U.S. Environmental Protection Agency launched Energy Star, a voluntary labeling program that is designed to promote and recognize the energy efficiency in monitors, climate control equipment, and other technologies. This resulted in the widespread adoption of sleep mode among consumer electronics. Concurrently, the Swedish organization TCO Development launched the TCO Certification program to promote low magnetic and electrical emissions from CRT-based computer displays; this program was later expanded to include criteria on energy consumption, ergonomics, and the use of hazardous materials in construction.

Regulations and Industry Initiatives

The Organisation for Economic Co-operation and Development (OECD) has published a survey of over 90 government and industry initiatives on "Green ICTs", i.e. information and communication technologies, the environment and climate change. The report concludes that initiatives tend to concentrate on the greening ICTs themselves rather than on their actual implementation to tackle global warming and environmental degradation. In general, only 20% of initiatives have measurable targets, with government programs tending to include targets more frequently than business associations.

Government

Many governmental agencies have continued to implement standards and regulations that encourage green computing. The Energy Star program was revised in October 2006 to include stricter efficiency requirements for computer equipment, along with a tiered ranking system for approved products.

By 2008, 26 US states established statewide recycling programs for obsolete computers and consumer electronics equipment. The statutes either impose an "advance recovery fee" for each unit sold at retail or require the manufacturers to reclaim the equipment at disposal.

In 2010, the American Recovery and Reinvestment Act (ARRA) was signed into legislation by President Obama. The bill allocated over $90 billion to be invested in green initiatives (renewable energy, smart grids, energy efficiency, etc.) In January 2010, the U.S. Energy Department granted $47 million of the ARRA money towards projects that aim to improve the energy efficiency of data centers. The projects provided research to optimize data center hardware and software, improve power supply chain, and data center cooling technologies.

Industry

- Climate Savers Computing Initiative (CSCI) is an effort to reduce the electric power consumption of PCs in active and inactive states. The CSCI provides a catalog of green products from its member organizations, and information for reducing PC power consumption. It was started on 2007-06-12. The name stems from the World Wildlife Fund's Climate Savers program, which was launched in 1999. The WWF is also a member of the Computing Initiative.

- The Green Electronics Council offers the Electronic Product Environmental Assessment Tool (EPEAT) to assist in the purchase of "greener" computing systems. The Council evaluates computing equipment on 51 criteria - 23 required and 28 optional - that measure a product's efficiency and sustainability attributes. Products are rated Gold, Silver, or Bronze, depending on how many optional criteria they meet. On 2007-01-24, President George W. Bush issued Executive Order 13423, which requires all United States Federal agencies to use EPEAT when purchasing computer systems.

- The Green Grid is a global consortium dedicated to advancing energy efficiency in data centers and business computing ecosystems. It was founded in February 2007 by several key companies in the industry – AMD, APC, Dell, HP, IBM, Intel, Microsoft, Rackable Systems, SprayCool, Sun Microsystems and VMware. The Green Grid has since grown to hundreds of members, including end-users and government organizations, all focused on improving Data Center Infrastructure Efficiency (DCIE).

- The Green500 list rates supercomputers by energy efficiency (megaflops/watt), encouraging a focus on efficiency rather than absolute performance.

- Green Comm Challenge is an organization that promotes the development of energy conservation technology and practices in the field of Information and Communications Technology (ICT).

- The Transaction Processing Performance Council (TPC) Energy specification augments existing TPC benchmarks by allowing optional publications of energy metrics alongside performance results.

- SPECpower is the first industry standard benchmark that measures power consumption in relation to performance for server-class computers. Other benchmarks which measure energy efficiency include SPECweb, SPECvirt, and VMmark.

Approaches

Modern IT systems rely upon a complicated mix of people, networks, and hardware; as such, a green computing initiative must cover all of these areas as well. A solution may also need to address end user satisfaction, management restructuring, regulatory compliance, and return on investment (ROI). There are also considerable fiscal motivations for companies to take control of their own power consumption; "of the power management tools available, one of the most powerful may still be simple, plain, common sense."

Product Longevity

Gartner maintains that the PC manufacturing process accounts for 70% of the natural resources used in the life cycle of a PC. More recently, Fujitsu released a Life Cycle Assessment (LCA) of a desktop that show that manufacturing and end of life accounts for the majority of this desktop's ecological footprint. Therefore, the biggest contribution to green computing usually is to prolong the equipment's lifetime. Another report from Gartner recommends to "Look for product longevity, including upgradability and modularity." For instance, manufacturing a new PC makes a far bigger ecological footprint than manufacturing a new RAM module to upgrade an existing one.

Data Center Design

Data center facilities are heavy consumers of energy, accounting for between 1.1% and 1.5% of the world's total energy use in 2010. The U.S. Department of Energy estimates that data center facilities consume up to 100 to 200 times more energy than standard office buildings.

Energy efficient data center design should address all of the energy use aspects included in a data center: from the IT equipment to the HVAC(Heating, ventilation and air conditioning) equipment to the actual location, configuration and construction of the building.

The U.S. Department of Energy specifies five primary areas on which to focus energy efficient data center design best practices:

- Information technology (IT) systems;

- Environmental conditions;

- Air management;

- Cooling systems;

- Electrical systems.

Additional energy efficient design opportunities specified by the U.S. Department of Energy include on-site electrical generation and recycling of waste heat.

Energy efficient data center design should help to better utilize a data center's space, and increase performance and efficiency.

In 2018, three new US Patents make use of facilities design to simultaneously cool and produce electrical power by use of internal and external waste heat. The three patents use silo design for stimulating use internal waste heat, while the recirculation of the air cooling the silo's computing racks. US Patent 9,510,486, uses the recirculating air for power generation, while sister patent, US Patent 9,907,213, forces the recirculation of the same air, and sister patent, US Patent 10,020,436, uses thermal differences in temperature resulting in negative power usage effectiveness. Negative power usage effectiveness, makes use of extreme differences between temperatures at times running the computing facilities, that they would run only from external sources other than the power use for computing.

Software and Deployment Optimization

Algorithmic Efficiency

The efficiency of algorithms affects the amount of computer resources required for any given computing function and there are many efficiency trade-offs in writing programs. Algorithm changes, such as switching from a slow (e.g. linear) search algorithm to a fast (e.g. hashed or indexed) search algorithm can reduce resource usage for a given task from substantial to close to zero. In 2009, a study by a physicist at Harvard estimated that the average Google search released 7 grams of carbon dioxide (CO_2). However, Google disputed this figure, arguing instead that a typical search produced only 0.2 grams of CO_2.

Resource Allocation

Algorithms can also be used to route data to data centers where electricity is less expensive. Researchers from MIT, Carnegie Mellon University, and Akamai have tested an energy allocation algorithm that successfully routes traffic to the location with the cheapest energy costs. The researchers project up to a 40 percent savings on energy costs if their proposed algorithm were to be deployed. However, this approach does not actually reduce the amount of energy being used; it reduces only the cost to the company using it. Nonetheless, a similar strategy could be used to direct traffic to rely on energy that is produced in a more environmentally friendly or efficient way. A similar approach has also been used to cut energy usage by routing traffic away from data centers experiencing warm weather; this allows computers to be shut down to avoid using air conditioning.

Larger server centers are sometimes located where energy and land are inexpensive and readily available. Local availability of renewable energy, climate that allows outside air to be used for cooling, or locating them where the heat they produce may be used for other purposes could be factors in green siting decisions.

Approaches to actually reduce the energy consumption of network devices by proper network/device management techniques are surveyed in. The approches are grouped into 4 main strategies, namely: (i) Adaptive Link Rate (ALR), (ii) Interface Proxying, (iii) Energy Aware Infrastructure, and (iv) Max Energy Aware Applications.

Virtualizing

Computer virtualization refers to the abstraction of computer resources, such as the process of running two or more logical computer systems on one set of physical hardware. The concept originated with the IBM mainframe operating systems of the 1960s, but was commercialized for x86-compatible computers only in the 1990s. With virtualization, a system administrator could combine several physical systems into virtual machines on one single, powerful system, thereby conserving resources by removing need for the original hardware and reducing power and cooling consumption. Virtualization can assist in distributing work so that servers are either busy or put in a low-power sleep state. Several commercial companies and open-source projects now offer software packages to enable a transition to virtual computing. Intel Corporation and AMD have also built proprietary virtualization enhancements to the x86 instruction set into each of their CPU product lines, in order to facilitate virtual computing.

New virtual technologies, such as operating-system-level virtualization can also be used to reduce energy consumption. These technologies make a more efficient use of resources, thus reducing energy consumption by design. Also, the consolidation of virtualized technologies is more efficient than the one done in virtual machines, so more services can be deployed in the same physical machine, reducing the amount of hardware needed.

Terminal Servers

Terminal servers have also been used in green computing. When using the system, users at a terminal connect to a central server; all of the actual computing is done on the server, but the end user experiences the operating system on the terminal. These can be combined with thin clients, which use up to 1/8 the amount of energy of a normal workstation, resulting in a decrease of energy costs and consumption. There has been an increase in using terminal services with thin clients to create virtual labs. Examples of terminal server software include Terminal Services for Windows and the Linux Terminal Server Project (LTSP) for the Linux operating system. Software-based remote desktop clients such as Windows Remote Desktop and RealVNC can provide similar thin-client functions when run on low power, commodity hardware that connects to a server.

Power Management

The Advanced Configuration and Power Interface (ACPI), an open industry standard, allows an operating system to directly control the power-saving aspects of its underlying hardware. This allows a system to automatically turn off components such as monitors and hard drives after set periods of inactivity. In addition, a system may hibernate, when most components (including the CPU and the system RAM) are turned off. ACPI is a successor to an earlier Intel-Microsoft standard called Advanced Power Management, which allows a computer's BIOS to control power management functions.

Some programs allow the user to manually adjust the voltages supplied to the CPU, which reduces both the amount of heat produced and electricity consumed. This process is called undervolting.

Some CPUs can automatically undervolt the processor, depending on the workload; this technology is called "SpeedStep" on Intel processors, "PowerNow"/"Cool'n'Quiet" on AMD chips, Long-Haul on VIA CPUs, and LongRun with Transmeta processors.

Data Center Power

Data centers, which have been criticized for their extraordinarily high energy demand, are a primary focus for proponents of green computing. According to a Greenpeace study, data centers represent 21% of the electricity consumed by the IT sector, which is about 382 billion kWh a year.

Data centers can potentially improve their energy and space efficiency through techniques such as storage consolidation and virtualization. Many organizations are aiming to eliminate underutilized servers, which results in lower energy usage. The first step toward this aim will be training of data center administrators. The U.S. federal government has set a minimum 10% reduction target for data center energy usage by 2011. With the aid of a self-styled ultraefficient evaporative cooling technology, Google Inc. has been able to reduce its energy consumption to 50% of that of the industry average.

Operating System Support

Microsoft Windows, has included limited PC power management features since Windows 95. These initially provided for stand-by (suspend-to-RAM) and a monitor low power state. Further iterations of Windows added hibernate (suspend-to-disk) and support for the ACPI standard. Windows 2000 was the first NT-based operating system to include power management. This required major changes to the underlying operating system architecture and a new hardware driver model. Windows 2000 also introduced Group Policy, a technology that allowed administrators to centrally configure most Windows features. However, power management was not one of those features. This is probably because the power management settings design relied upon a connected set of per-user and per-machine binary registry values, effectively leaving it up to each user to configure their own power management settings.

This approach, which is not compatible with Windows Group Policy, was repeated in Windows XP. The reasons for this design decision by Microsoft are not known, and it has resulted in heavy criticism. Microsoft significantly improved this in Windows Vista by redesigning the power management system to allow basic configuration by Group Policy. The support offered is limited to a single per-computer policy. The most recent release, Windows 7 retains these limitations but does include refinements for timer coalescing, processor power management, and display panel brightness. The most significant change in Windows 7 is in the user experience. The prominence of the default High Performance power plan has been reduced with the aim of encouraging users to save power.

There is a significant market in third-party PC power management software offering features beyond those present in the Windows operating system. available. Most products offer Active Directory integration and per-user/per-machine settings with the more advanced offering multiple power plans, scheduled power plans, anti-insomnia features and enterprise power usage reporting. Notable vendors include 1E NightWatchman, Data Synergy PowerMAN (Software), Faronics Power Save, Verdiem SURVEYOR and EnviProt Auto Shutdown Manager.

Linux systems started to provide laptop-optimized power-management in 2005, with power-management options being mainstream since 2009.

Power Supply

Desktop computer power supplies are in general 70–75% efficient, dissipating the remaining energy as heat. A certification program called 80 Plus certifies PSUs that are at least 80% efficient; typically these models are drop-in replacements for older, less efficient PSUs of the same form factor. As of July 20, 2007, all new Energy Star 4.0-certified desktop PSUs must be at least 80% efficient.

Storage

Smaller form factor (e.g., 2.5 inch) hard disk drives often consume less power per gigabyte than physically larger drives. Unlike hard disk drives, solid-state drives store data in flash memory or DRAM. With no moving parts, power consumption may be reduced somewhat for low-capacity flash-based devices.

In a recent case study, Fusion-io, manufacturer of solid state storage devices, managed to reduce the energy use and operating costs of MySpace data centers by 80% while increasing performance speeds beyond that which had been attainable via multiple hard disk drives in Raid 0. In response, MySpace was able to retire several of their servers.

As hard drive prices have fallen, storage farms have tended to increase in capacity to make more data available online. This includes archival and backup data that would formerly have been saved on tape or other offline storage. The increase in online storage has increased power consumption. Reducing the power consumed by large storage arrays, while still providing the benefits of online storage, is a subject of ongoing research.

Video Card

A fast GPU may be the largest power consumer in a computer.

Energy-efficient display options include:

- No video card - Use a shared terminal, shared thin client, or desktop sharing software if display required.

- Use motherboard video output - Typically low 3D performance and low power.

- Select a GPU based on low idle power, average wattage, or performance per watt.

Display

Unlike other display technologies, electronic paper does not use any power while displaying an image. CRT monitors typically use more power than LCD monitors. They also contain significant amounts of lead. LCD monitors typically use a cold-cathode fluorescent bulb to provide light for the display. Some newer displays use an array of light-emitting diodes (LEDs) in place of the fluorescent bulb, which reduces the amount of electricity used by the display. Fluorescent back-lights also contain mercury, whereas LED back-lights do not.

Materials Recycling

Recycling computing equipment can keep harmful materials such as lead, mercury, and hexavalent chromium out of landfills, and can also replace equipment that otherwise would need to be manufactured, saving further energy and emissions. Computer systems that have outlived their particular function can be re-purposed, or donated to various charities and non-profit organizations. However, many charities have recently imposed minimum system requirements for donated equipment. Additionally, parts from outdated systems may be salvaged and recycled through certain retail outlets and municipal or private recycling centers. Computing supplies, such as printer cartridges, paper, and batteries may be recycled as well.

A drawback to many of these schemes is that computers gathered through recycling drives are often shipped to developing countries where environmental standards are less strict than in North America and Europe. The Silicon Valley Toxics Coalition estimates that 80% of the post-consumer e-waste collected for recycling is shipped abroad to countries such as China and Pakistan.

In 2011, the collection rate of e-waste is still very low, even in the most ecology-responsible countries like France. In this country, e-waste collection is still at a 14% annual rate between electronic equipment sold and e-waste collected for 2006 to 2009.

The recycling of old computers raises an important privacy issue. The old storage devices still hold private information, such as emails, passwords, and credit card numbers, which can be recovered simply by someone's using software available freely on the Internet. Deletion of a file does not actually remove the file from the hard drive. Before recycling a computer, users should remove the hard drive, or hard drives if there is more than one, and physically destroy it or store it somewhere safe. There are some authorized hardware recycling companies to whom the computer may be given for recycling, and they typically sign a non-disclosure agreement.

Cloud Computing

Cloud computing addresses two major ICT challenges related to Green computing – energy usage and resource consumption. Virtualization, Dynamic provisioning environment, multi-tenancy, green data center approaches are enabling cloud computing to lower carbon emissions and energy usage up to a great extent. Large enterprises and small businesses can reduce their direct energy consumption and carbon emissions by up to 30% and 90% respectively by moving certain on-premises applications into the cloud. One common example includes Online shopping that helps people purchase products and services over the Internet without requiring them to drive and waste fuel to reach out to the physical shop, which, in turn, reduces greenhouse gas emission related to travel.

Edge Computing

New technologies such as Edge and Fog computing are a solution to reducing energy consumption. These technologies allow redistributing computation near the use, thus reducing energy costs in the network. Furthermore, having smaller data centers, the energy used in operations such as refrigerating and maintenance gets largely reduced.

Telecommuting

Teleconferencing and telepresence technologies are often implemented in green computing initiatives. The advantages are many; increased worker satisfaction, reduction of greenhouse gas emissions related to travel, and increased profit margins as a result of lower overhead costs for office space, heat, lighting, etc. The savings are significant; the average annual energy consumption for U.S. office buildings is over 23 kilowatt hours per square foot, with heat, air conditioning and lighting accounting for 70% of all energy consumed. Other related initiatives, such as Hoteling, reduce the square footage per employee as workers reserve space only when they need it. Many types of jobs, such as sales, consulting, and field service, integrate well with this technique.

Voice over IP (VoIP) reduces the telephony wiring infrastructure by sharing the existing Ethernet copper. VoIP and phone extension mobility also made hot desking more practical.

Telecommunication Network Devices Energy Indices

The information and communication technologies (ICTs) energy consumption, in the USA and worldwide, has been estimated respectively at 9.4% and 5.3% of the total electricity produced. The energy consumption of ICTs is today significant even when compared with other industries. Some study tried to identify the key energy indices that allow a relevant comparison between different devices (network elements). This analysis was focused on how to optimise device and network consumption for carrier telecommunication by itself. The target was to allow an immediate perception of the relationship between the network technology and the environmental effect.

Supercomputers

The inaugural Green500 list was announced on November 15, 2007 at SC|07. As a complement to the TOP500, the unveiling of the Green500 ushered in a new era where supercomputers can be compared by performance-per-watt.

The TSUBAME-KFC-GSIC Center by Tokyo Institute of Technology, Made in Japan was with a great advantage to the second, the Top 1 Supercomputer in the World with 4,503.17 MFLOPS/W and 27.78 Total Power (kW)++.

Today a new supercomputer, L-CSC from the GSI Helmholtz Center, Made in Germany emerged as the most energy-efficient (or greenest) supercomputer in the world. The L-CSC cluster was the first and only supercomputer on the list to surpass 5 gigaflops/watt (billions of operations per second per watt). L-CSC is a heterogeneous supercomputer that is powered by Dual Intel Xeon E5-260 and GPU accelerators, namely AMD FirePro S9150 GPUs. It marks the first time that a supercomputer using AMD GPUs has held the top spot. Each server has a memory of 256 gigabytes. Connected, the server via an Infiniband FDR network.

Passive House

Passive house is a building standard that is truly energy efficient, comfortable, affordable and ecological at the same time.

Passive House is not a brand name, but a construction concept that can be applied by anyone and that has stood the test of practice.

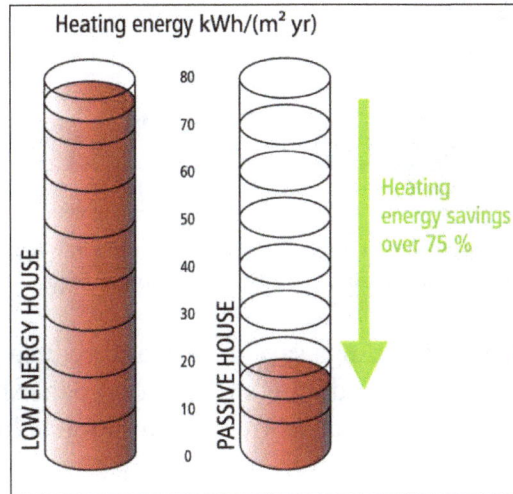

Yet, a Passive House is more than just a low-energy building:

- Passive House buildings allow for heating and cooling related energy savings of up to 90% compared with typical building stock and over 75% compared with average new builds. In terms of heating oil, Passive House buildings use less than 1.5 litres per square meter of living space per year – far less than typical low-energy buildings. Similar energy savings have been demonstrated in warm climates where buildings require more energy for cooling than for heating.

- Passive House buildings are also praised for their high level of comfort. They use energy sources inside the building such as the body heat from the residents or solar heat entering the building – making heating a lot easier.

- Appropriate windows with good insulation and a building shell consisting of good insulated exterior walls, roof and floor slab keep the heat during winter in the house – and keep it out during summer.

- A ventilation system consistently supplies fresh air making for superior air quality without causing any unpleasant draughts. This is a guarantee for low Radon levels and improving the health conditions. A highly efficient heat recovery unit allows for the heat contained in the exhaust air to be re-used.

The vast energy savings in Passive House buildings are achieved by using especially energy efficient building components and a quality ventilation system: There is absolutely no cutting back on comfort; instead the level of comfort is considerably increase.

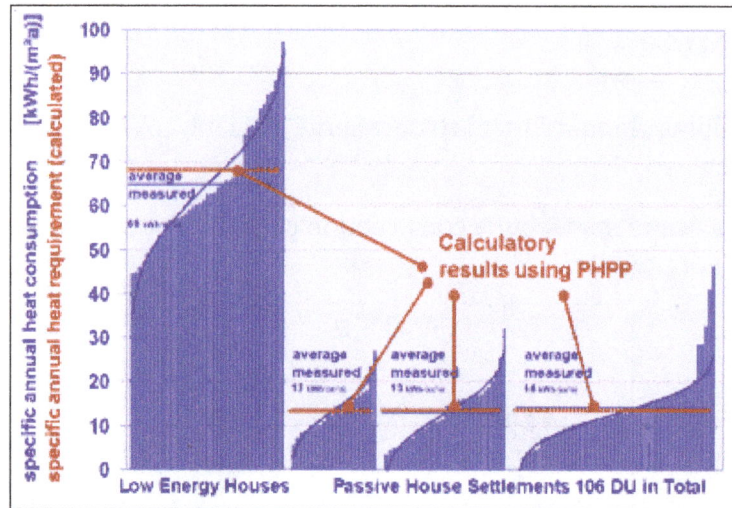

Heat consumption measured in 4 residential estates: One low-energy estate (left) and three Passive House estates. No "perfomance gap" here.

In the above image, Passive House buildings save energy and reduce greenhouse gases - not just a little but a great deal. And these savings do not only exist on paper but also in real life - passive house buildings do deliver. This diagram shows the consumption values measured in low-energy houses and in Passive House estates.

Comfort

The Passive House Standard offers a new level of quality pairing a maximum level of comfort both during cold and warm months with reasonable construction costs – something that is repeatedly confirmed by Passive House residents.

Quality

In the above image, Passive House buildings are praised for their efficiency due to their high level of insulation and their airtight design. Another important principle is "thermal bridge free design": the insulation is applied without any "weak spots" around the whole building so as to eliminate cold corners as well as excessive heat losses. This method is another essential principle assuring a high level of quality and comfort in Passive House buildings while preventing damages due to moisture build up.

Ecology/Sustainability

Passive House buildings are eco-friendly by definition: They use extremely little primary energy, leaving sufficient energy resources for all future generations without causing any environmental damage. The additional energy required for their construction (embodied energy) is rather insignificant compared with the energy they save later on. This seems so obvious that there is no immediate need for additional illustrations. It is rather worth mentioning though, that the Passive House standard provides this level of sustainability for anyone wishing to build a new construction or renovating an older one at an affordable price – A contribution to protecting the environment. Be aware, that the principles are all published and the design tools are made available for all architects.

Affordability

Are Passive House buildings a good investment? Passive House buildings not only save money over the long term, but are surprisingly affordable to begin with. The investment in higher quality building components required by the Passive House Standard is mitigated by the elimination of expensive heating and cooling systems. Additional financial support increasingly available in many countries makes building a Passive House all the more feasible.

Measurement Results

Measurements carried out in 114 Passive House apartments which were part of the CEPHEUS project showed average savings of approx. 90%. In other words, the Passive House is a "factor 10 house" which only uses one tenth of the energy used by average houses.The passive house concept delivers - the savings are real, there is no performance gap.

Versatility

Any competent architect can design a Passive House. By combining individual measures any new building anywhere in the world can be designed to reach the Passive House Standard. The versatile Passive House Standard is also increasingly being used for non-residential buildings such as administrative buildings and schools. Education on the design of passive house buildings is available on a global level with a lot of different professional trainers.

Retrofits

The Passive House Standard can also be achieved in retrofits using Passive House components.

The International Passive House Open Days takes place once a year. At this event, hundreds of Passive House residents open their doors to anyone interested in getting a hands-on experience of living in a Passive House. The International Passive House Days are put on by the International Passive House Association (iPHA) and its German affiliate, IG Passivhaus.

Working

The Passive House is the world leading standard in energy-efficient construction: A Passive House requires as little as 10 percent of the energy used by typical Central European buildings – meaning an energy savings of up to 90 percent. Owners of Passive Houses are barely concerned with increasing energy prices.

- Passive Houses require less than 15 kWh/(m^{2}yr) for heating or cooling (relating to the living space).

- The heating/cooling load is limited to a maximum of 10 W/m^2.

- Conventional Primary energy use may not exceed 120 kWh/(m^{2}a) - but the future is renewable energy supply (PER) with no more than 60 kWh/(m^{2}a). This is easy to accomplish with passive houses.

- Passive Houses must be airtight with air change rates being limited to n50 = 0.6/h.

- In warmer climates and/or during summer months, excessive temperatures may not occur more than 10 % of the time.

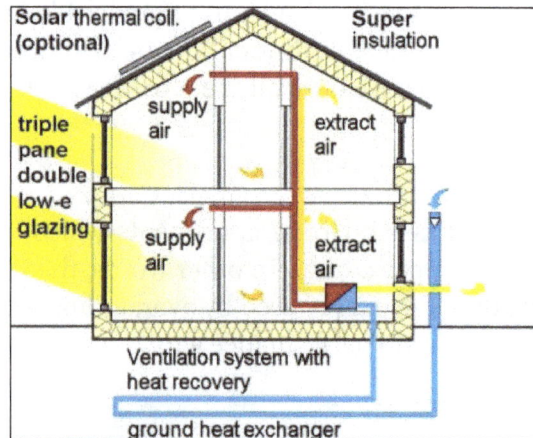

The Passive House is a sustainable construction concept that provides for affordable, high-quality buildings as well as comfortable, healthy living conditions. And its principles are quite easy to understand:

- As newer buildings are increasingly airtight, ventilation through joints and cracks alone is not sufficient to provide for fresh indoor air. Opening the windows as recommended won't do the job either. Fresh air is not merely a matter of comfort but a necessity for healthy living - Indoor Air Quality (IAQ) is the basic performance goal. Ventilation systems are therefore the key technology for all future residential buildings and retrofits.

- Even though ventilation systems do require an extra investment to begin with they will end up saving considerable amounts of energy costs, provided that they are highly efficient systems. Passive House quality ventilation systems will reduce the operating costs of any building.

- This is where the Passive House concept comes in: As large amounts of fresh outdoor air need to be supplied to the building anyway, why not use this air for heating? - Without any extra amounts of air, without any recirculation of air, without any inconvenient noise or drafts? This way the ventilation system pays off twice.

- This "supply air heating" concept only works in appropriately insulated buildings – that is in Passive Houses. In expert terms: The transmission and infiltration heating load must be less than 10 W/m² to make sure that the required heat can be provided by the supply air.

Passive House Characteristics

- The Passive House concept doesn't only work on paper – Built Passive House examples around the world prove that it also works in real buildings.

- The Passive House concept has been proven to works in real life: Passive House – measurement results.

Several thousand Passive House dwellings have been monitored with respect to air quality, thermal comfort, energy consumption and construction as well as operating costs. These results have been published. The results show that the concept fully lives up to its promises: An improved level of comfort paired with extremely low energy consumption - sustainably low.

- Passive Houses are affordable. Building professionals in several countries with different climatic conditions and building tradiations have shown: It is possible to develop the Passive House Standard based on the existing experiences and knowledge in the building sector. All it takes is: specific know how and best practice building components (i.e. windows, heat recovery units).

- Passive Houses offer a maximum level of comfort. For some 40 years the meaning of superior termal comfort has been well established on a scientific level: It was the publication of Ole Fanger in 1972: "Thermal Comfort" whose results became the basis for modern international standards on thermal comfort such as ISO 7730. The Passive House concept is based on a thorough analysis of how to achieve superior comfort levels with mainly passive components in different circumstances of the natural environment.

- Passive Houses are sustainable.

- Passives Houses mostly benefit regional manufacturers - everybody is invited to contribute. Products for Passive Houses are best practise products all around the world: Minimal U-values, superior window energy performance, maximum heat recovery rates. These products are mainly produced by small and medium enterprises on a local and regional level. This is adequate for passive components, because the resources are available anywhere. Insulation materials e.g. can be produced from a lot of very different resources - the main material always being just air included in small spaces, moving very slowly. The structural material can be wool, straw, wood fibers, paper, mineral wool, several types of plastics, foamed calcium silicate, foamed glass.

Zero Energy Building

Zero-energy building (ZEB), also called net zero-energy building is any building or construction characterized by zero net energy consumption and zero carbon emissions calculated over a period of time. Zero-energy buildings (ZEBs) usually use less energy than traditional buildings as well as generate their own energy on-site to use in the building; hence, many are independent from the national (electricity) grid. ZEBs have emerged in response to stringent environmental standards, both regulatory and voluntary, introduced to address increasingly significant environmental issues such as climate change, natural resource conservation, pollution, ecology, and population.

Many people in developing countries (and elsewhere) already live in zero-energy buildings out of necessity, including huts, tents, and caves exposed to temperature extremes and without access to electricity. The notion of a "zero-energy building" in a modern sense has been discussed since the 1970s, prompted by the petroleum shocks of the decade and subsequent concerns about the consequences of fossil fuel dependency. Definitions of ZEBs vary from those related to net energy inputs versus outputs to those that balance the financial costs of energy use with the costs associated with equipment used in the structure for energy production—from photovoltaics (solar cells) and wind turbines, for example—combined with the benefits associated with exporting energy generated by the structure. The energy in a building can be measured in many ways (e.g., cost, energy, or carbon emissions), and different views exist on the relative importance of energy production and energy conservation in achieving a net energy balance.

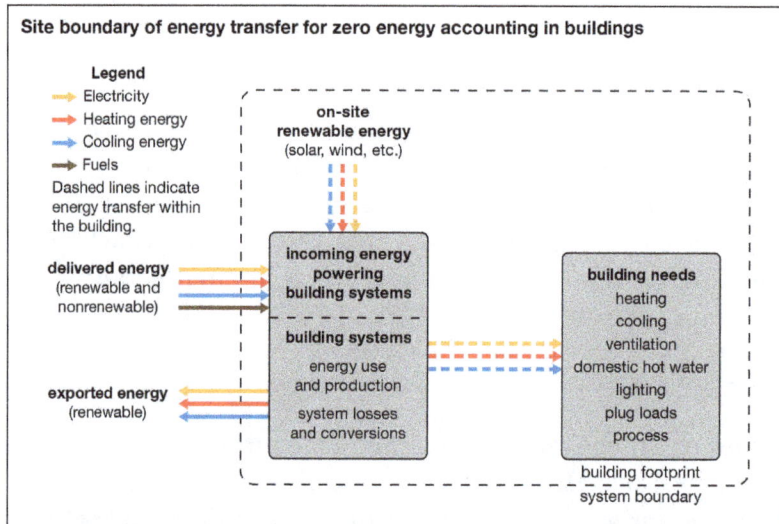

Zero-energy building: Site boundary of energy transfer.

Zero-energy buildings (ZEBs) usually use less energy than traditional buildings as well as generate their own energy on-site to use in the building; hence, many are independent from the national (electricity) grid.

ZEB Energy Generation

ZEBs need to produce their own energy on site to meet their electricity and heating or cooling needs. Various microgeneration technologies may be used to provide heat and electricity to the building, including the following:

- Solar (solar hot water, photovoltaics [PV]).

- Wind (wind turbines).

- Biomass (heaters and stoves, boilers, and community heating schemes).

- Combined heat and power (CHP) and micro-CHP for use with natural gas, biomass, sewerage gas, and other biogases.

- Community heating (including utilizing waste heat from large-scale power generation).

- Heat pumps (air source [ASHP] and ground source [GSHP] and geothermal heating systems).

- Water (small-scale hydropower).

- Other (including fuel cells using hydrogen generated from any of the above renewable sources).

Many homebuilders have serious concerns about whether microgeneration and renewable energy technologies can deliver the energy generation requirements to produce adequate working, cost-effective ZEBs. Builders fear that owners and occupiers may not accept the required new technologies and could choose to retrofit energy-intensive appliances and systems, which would ultimately undermine the zero-energy objectives. There are further concerns that failure to maintain the new systems and technologies adequately may expose owners and occupiers to health and safety risks.

Smart Grid

The Smart Grid, refers to the electric grid, a network of transmission lines, substations, transformers and more that deliver electricity from the power plant to your home or business. It's what you plug into when you flip on your light switch or power up your computer. Our current electric grid was built in the 1890s and improved upon as technology advanced through each decade. Today, it consists of more than 9,200 electric generating units with more than 1 million megawatts of generating capacity connected to more than 300,000 miles of transmission lines. Although the electric grid is considered an engineering marvel, we are stretching its patchwork nature to its capacity. To move forward, we need a new kind of electric grid, one that is built from the bottom up to handle the groundswell of digital and computerized equipment and technology dependent on it—and one that can automate and manage the increasing complexity and needs of electricity in the 21st Century.

In short, the digital technology that allows for two-way communication between the utility and its customers, and the sensing along the transmission lines is what makes the grid smart. Like the Internet, the Smart Grid will consist of controls, computers, automation, and new technologies and equipment working together, but in this case, these technologies will work with the electrical grid to respond digitally to our quickly changing electric demand.

Functions of a Smart Grid

The Smart Grid represents an unprecedented opportunity to move the energy industry into a new era of reliability, availability, and efficiency that will contribute to our economic and environmental health. During the transition period, it will be critical to carry out testing, technology improvements, consumer education, development of standards and regulations, and information sharing between projects to ensure that the benefits we envision from the Smart Grid become a reality. The benefits associated with the Smart Grid include:

- More efficient transmission of electricity.

- Quicker restoration of electricity after power disturbances.

- Reduced operations and management costs for utilities, and ultimately lower power costs for consumers.

- Reduced peak demand, which will also help lower electricity rates.

- Increased integration of large-scale renewable energy systems.

- Better integration of customer-owner power generation systems, including renewable energy systems.

- Improved security.

Today, an electricity disruption such as a blackout can have a domino effect—a series of failures that can affect banking, communications, traffic, and security. This is a particular threat in the winter, when homeowners can be left without heat. A smarter grid will add resiliency to our electric power System and make it better prepared to address emergencies such as severe storms, earthquakes, large solar flares, and terrorist attacks. Because of its two-way interactive capacity, the Smart Grid will allow for automatic rerouting when equipment fails or outages occur. This will minimize outages and minimize the effects when they do happen. When a power outage occurs, Smart Grid technologies will detect and isolate the outages, containing them before they become large-scale blackouts. The new technologies will also help ensure that electricity recovery resumes quickly and strategically after an emergency—routing electricity to emergency services first, for example. In addition, the Smart Grid will take greater advantage of customer-owned power generators to produce power when it is not available from utilities. By combining these "distributed generation" resources, a community could keep its health center, police department, traffic lights, phone System, and grocery store operating during emergencies. In addition, the Smart Grid is a way to address an aging energy infrastructure that needs to be upgraded or replaced. It's a way to address energy efficiency, to bring increased awareness to consumers about the connection between electricity use and the environment. And it's a way to bring increased national security to our energy System—drawing on greater amounts of home-grown electricity that is more resistant to natural disasters and attack.

Giving Consumers Control

The Smart Grid is not just about utilities and technologies; it is about giving you the information and tools you need to make choices about your energy use. If you already manage activities such as personal banking from your home computer, imagine managing your electricity in a similar way. A smarter grid will enable an unprecedented level of consumer participation. For example, you will no longer have to wait for your monthly statement to know how much electricity you use. With a smarter grid, you can have a clear and timely picture of it. "Smart meters," and other mechanisms, will allow you to see how much electricity you use, when you use it, and its cost. Combined with real-time pricing, this will allow you to save money by using less power when electricity is most expensive. While the potential benefits of the Smart Grid are usually discussed in terms of economics, national security, and renewable energy goals, the Smart Grid has the potential to help you save money by helping you to manage your electricity use and choose the best times to purchase electricity. And you can save even more by generating your own power.

Building and Testing the Smart Grid

The Smart Grid will consist of millions of pieces and parts—controls, computers, power lines, and new technologies and equipment. It will take some time for all the technologies to be perfected, equipment installed, and systems tested before it comes fully on line. And it won't happen all at once—the Smart Grid is evolving, piece by piece, over the next decade or so. Once mature, the Smart Grid will likely bring the same kind of transformation that the Internet has already brought to the way we live, work, play, and learn.

References

- "Heat Island Group Home Page". Lawrence Berkeley National Laboratory. 2000-08-30. Archived from the original on January 9, 2008. Retrieved 2008-01-19

- 80-green-building, environment-and-architecture: environment-ecology.com, Retrieved 29 March, 2019

- "BPI Certifications - Certifications for Skilled, Advanced Home Energy, Entry Level Practitioner, and Multi-Family Building professionals". Retrieved 30 June 2015

- Zero-energy-building, technology: britannica.com, Retrieved 15 March, 2019

- "Seattle City Light/Seattle Public Utilities Home Resources Profile". Seattle.gov. Retrieved 2012-07-26

- Alternative-energy-sources: renewableresourcescoalition.org, Retrieved 12 July, 2019

- "TCO takes the initiative in comparative product testing". May 3, 2008. Archived from the original on July 23, 2007. Retrieved May 3,2008

- Smart-grid, the-smart-grid: smartgrid.gov, Retrieved 10 June, 2019

- Jones, Ernesta (October 23, 2006). "EPA Announces New Computer Efficiency Requirements". U.S. EPA. Archived from the original on February 12, 2007. Retrieved September 18, 2007

- What-is-a-passive-house, basics: passipedia.org, Retrieved 12 August, 2019

- "Big Rig Building Boom". Rigzone.com. 2006-04-13. Archived from the original on 2007-10-21. Retrieved 2008-01-18

Sustainable Energy | 5

- **Energy Hierarchy**
- **Sustainable Energy**
- **Energy Conservation**
- **Energy Storage**

Sustainable energy is a practice that focuses on fulfilling the present demands of energy without jeopardizing the ability of future generation to meet their needs. The efforts which are made in order to reduce the consumption of energy by lessening the usage of a particular energy service are termed as energy conservation. The diverse aspects of sustainable energy and energy conservation have been thoroughly discussed in this chapter.

Energy Hierarchy

Saving Energy for the motive of sustainable development is one of the most sought after issue that is emphasized upon by the modern day governments. For this very reason there exists a system of priority in the energy hierarchy that is meant to be understood by various corporate houses to minimize the unnecessary depletion of resources and subsequently promote conservation of energy. These hierarchies in energy have emerged in the past two decades with the coming in of innovations in the markets and therefore the increased costs of resources. Added to it was the new emerging global concern when it was realized that utilization of energy became an evident contributor and cause for global extinction if not utilized properly.

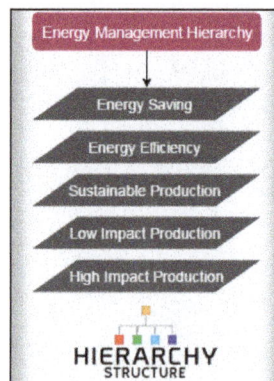

Therefore here is the hierarchy of energy that should be considered in priority order to ensure a long term success and efficiency in the energy production.

- Energy saving should be the top most priority for all corporate houses, it can be achieved through preventing unnecessary utilization of energy resources which is possible by reducing utilization of heavy volt machines and appliances or cutting down on journeys that are not required.

- Energy efficiency to ensure that energy is produced and consumed properly through conversion. This is next to saving and has very recently developed as a concept due to the increasing awareness among people and a sudden rise in prices of energy.

- Sustainable production has also emerged in the hierarchies with the realization of the need to save resources of energy. It is there obtained through renewable resources, that either do not exhaust such as wind, sunlight, rainwater, waves, earth, etc or the bio energy that is converted through combustion and decay of gases like bio-methane, liquids like bio-fuels or most commonly used solid waste or solids like energy crop and wood.

Though these are highly recommended, yet are inaccessible to a major portion due to the lack of proper knowledge and technology to execute such energy productions in the most efficient manners.

- Low impact production is the next option available. It is not a very effective manner from the business point of view since at high cost there is less production but this option is still available as emits the least level of carbons in the air. Additionally they also have low carbon impact since they are not completely sustainable in nature. Some consider nuclear energy as one of the low impact sources.

- High impact production which is although highly preferred for business purposes due to its cost but comes last in the energy hierarchy as it has the worst impact on the environment. These include unsustainable sources like unabated fossils, highly radioactive nuclear emissions, etc. Energy utilization of such resources has to be minimized to the least.

The very first system of energy production introduced in industries and governments had three tiers; traditional production, production from renewable resources and energy efficiency. Consequently the concept was developed to be the current five tier system by various industries across the world.

Sustainable Energy

Sustainable energy is a form of energy that meet our today's demand of energy without putting them in danger of getting expired or depleted and can be used over and over again. Sustainable energy should be widely encouraged as it do not cause any harm to the environment and is available widely free of cost. All renewable energy sources like solar, wind, geothermal, hydropower and ocean energy are sustainable as they are stable and available in plenty.

Sun will continue to provide sunlight till we all are here on earth, heat caused by sun will continue to produce winds, earth will continue to produce heat from inside and will not cool down anytime soon, movement of earth, sun and moon will not stop and this will keep on producing tides and the process of evaporation will cause water to evaporate that will fall down in the form of rain or ice which will go through rivers or streams and merge in the oceans and can be used to produce energy through hydropower. This clearly states that all these renewable energy sources are sustainable and will continue to provide energy to the coming generations.

There are many forms of sustainable energy sources that can be incorporated by countries to stop the use of fossil fuels. Sustainable energy does not include any sources that are derived from fossil fuels or waste products. This energy is replenishable and helps us to reduce greenhouse gas emissions and causes no damage to the environment. If we are going to use fossil fuels at a steady rate, they will expire soon and cause adverse affect to our planet.

Fossil fuels are not considered as sustainable energy sources because they are limited, cause immense pollution by releasing harmful gases and are not available everywhere on earth. Fossil fuels normally include coal, oil and natural gas. Steps must be taken to reduce our dependency on fossil fuels as pose dangerous to environment. Most of the counties have already started taking steps to make use of alternative energy sources. As of today, around 20% of world's energy needs comes from renewable energy sources. Hydropower is the most common form of alternative energy used around the world.

Need for Sustainable Energy

During ancient times, wood, timber and waste products were the only major energy sources. In short, biomass was the only way to get energy. When more technology was developed, fossil fuels like coal, oil and natural gas were discovered. Fossil fuels proved boom to the mankind as they were widely available and could be harnessed easily. When these fossil fuels were started using extensively by all the countries across the globe, they led to degradation of environment. Coal and oil are two of the major sources that produce large amount of carbon dioxide in the air. This led to increase in global warming.

Also, few countries have hold on these valuable products which led to the rise in prices of these fuels. Now, with rising prices, increasing air pollution and risk of getting expired soon forced scientists to look out for some alternative or renewable energy sources. The need of the hour was to look for resources that are available widely, cause no pollution and are replenishable. Sustainable Energy, at that time came into the picture as it could meet our today's increasing demand of energy and also provide us with an option to make use of them in future also.

Types of Sustainable Energy

Sustainable energy are not just a part of renewable energy sources, they are also the sources of energy that can best be used to power homes and industries without any harmful effects being experienced. This is the sole reason why many people advice the use of these forms of energy in everyday life. It is because its effects to the environment are purely beneficial.

Solar Energy

Solar energy is the best form of sustainable energy. This energy manifests itself in tow forms. There is the light and the heat. Both of these forms are equally important to us in our day to day living and other forms of life. For instance, the plants need the light to grow and generate food while man needs the heat energy to maintain body temperature and power their homes and industries. This means that it is the greatest form of sustainable energy. It can be used two folds with greater results as needed. This only serves to generate confidence and ensure that we live the way we intended without causing further harm to the environment.

According to activists, it is the future of energy. Evidence of intensive use of this alternative energy source can be seen everywhere. There are many companies that are making solar panels to tap this energy for use at home or in the industries. Consequently, the energy is also being tapped for commercial purposes in many fields like powering of homes in power grids. All that one needs to do is to get hold of the solar panel and install it in the homes or commercial property. During the summer periods, you can cut down on your energy costs.

Wind Energy

Wind is a sustainable energy source. It is available naturally and can be tapped to produce vast amounts of power that can be used in many ways and places. For instance, sailors tap this energy to help the ship propel through its various directions to distant shores for trading. Nowadays, this energy sources is being commercialized. There are many companies that have invested heavily on power grids and windmills to tap into this energy source. The energy generated can be sold to other people to power their homes and industries. In the near future, sustainable energy like wind power will be a big industry and the fossil fuels exploration will have halted and no longer being used.

Geothermal Energy

Geothermal energy allows us fetch the energy from beneath the earth. This occurs by installing geothermal power stations that can use heat coming out from inside the earth and use it to generate electricity. The temperature below the earth around 10,000 meters is so high that it can used to boil water. Geothermal energy cannot be harnessed everywhere as high temperature is needed to produce steam that could move turbines. It can be harnessed in those areas that have high seismic activity and are prone to volcanoes. They are environment friendly and can produce energy throughout the day but their ability to produce energy at suitable regions restricts us from using it on a much wider scale.

Ocean Energy

There is massive size of oceans in this world. About 70% of the earth is covered with water. The potential that ocean energy has to generate power is much higher than any other source of energy. This sustainable energy allows us to harness it in 3 ways i.e. wave, tidal or ocean thermal energy conversion (OTEC). Tides have immense power which when effectively tapped can generate a lot of energy and can be used to power millions of homes. Waves produced at the oceans can be used by ocean thermal plants to convert the kinetic energy in waves to mechanical energy of turbines which can again converted to electrical energy through generators. Setting up of big plants at ocean may cause ecological imbalance and disturb aquatic life.

Biomass Energy

Biomass energy is produced by burning of wood, timber, landfills and municipal and agricultural waste. It is completely renewable and does not produce harmful gases like carbon dioxide which is primarily responsible for increase in global warming. Although, carbon dioxide is produced by burning these products but that is equally compensated when plants take this carbon dioxide and produce oxygen. It also helps to reduce landfills but are not as effective as fossil fuels.

Hydroelectric Power

On the other hand, there are the rivers or waterfalls whose energy of the moving water is captured that can turn turbines to generate power. This is commonly known as hydroelectric power. It is very common nowadays and it is powering most parts of the world and one of the biggest form of alternative energy currently being used. There are many companies and countries that are exporting this energy to other countries who unable to harness it on their own due to lack of the necessary resources or conditions. The energy is commonly transported in form of power lines to various parts of the country and even outside the country.

These are the three best case examples of sustainable energy forms that are projected to run the world in the near future. They are very sustainable and so not cause any environmental effects. Their inability to be depleted and lack of effect to the environmental makes them a perfect candidate to future energy needs.

Energy Conservation

Energy conservation is the effort made to reduce the consumption of energy by using less of an energy service. This can be achieved either by using energy more efficiently (using less energy for a constant service) or by reducing the amount of service used (for example, by driving less). Energy conservation is a part of the concept of eco-sufficiency. Energy conservation reduces the need for energy services and can result in increased environmental quality, national security, personal financial security and higher savings. It is at the top of the sustainable energy hierarchy. It also lowers energy costs by preventing future resource depletion.

Energy can be conserved by reducing wastage and losses, improving efficiency through technological upgrades and improved operation and maintenance. On a global level energy use can also be reduced by the stabilisation of population growth.

Energy can only be transformed from one form to other, such as heat energy to motive power in cars, or kinetic energy of water flow to electricity in hydroelectric power plants. However machines are required to transform energy from one form to other. The wear and friction of the components of these machine while running cause loss of quadrillions of BTU and $500 billions in industries only in USA. It is possible to minimize these losses by adopting green engineering practices to improve life cycle of the components.

Energy Tax

Some countries employ energy or carbon taxes to motivate energy users to reduce their consumption. Carbon taxes can force consumption to shift to nuclear power and other energy sources that carry different sets of environmental side effects and limitations. On the other hand, taxes on all energy consumption can reduce energy use across the board while reducing a broader array of environmental consequences arising from energy production. The state of California employs a tiered energy tax whereby every consumer receives a baseline energy allowance that carries a low tax. As usage increases above that baseline, the tax increases drastically. Such programs aim to protect poorer households while creating a larger tax burden for high energy consumers.

Building Design

Occupancy sensors can conserve energy by turning off appliances in unoccupied rooms.

One of the primary ways to improve energy conservation in buildings is to perform an energy audit. An energy audit is an inspection and analysis of energy use and flows for energy conservation in a building, process or system with an eye toward reducing energy input without negatively affecting output. This is normally accomplished by trained professionals and can be part of some of the national programs discussed above. Recent development of smartphone apps enables homeowners to complete relatively sophisticated energy audits themselves.

Building technologies and smart meters can allow energy users, both commercial and residential, to visualize the impact their energy use can have in their workplace or homes. Advanced real-time energy metering can help people save energy by their actions.

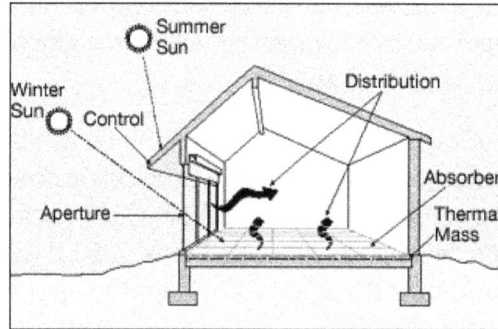

Elements of passive solar design, shown in a direct gain application.

In passive solar building design, windows, walls, and floors are made to collect, store, and distribute solar energy in the form of heat in the winter and reject solar heat in the summer. This is called passive solar design or climatic design because, unlike active solar heating systems, it does not involve the use of mechanical and electrical devices.

The key to designing a passive solar building is to best take advantage of the local climate. Elements to be considered include window placement and glazing type, thermal insulation, thermal mass, and shading. Passive solar design techniques can be applied most easily to new buildings, but existing buildings can be retrofitted.

Transportation

In the United States, suburban infrastructure evolved during an age of relatively easy access to fossil fuels, which has led to transportation-dependent systems of living. Zoning reforms that allow greater urban density as well as designs for walking and bicycling can greatly reduce energy consumed for transportation. The use of telecommuting by major corporations is a significant opportunity to conserve energy, as many Americans now work in service jobs that enable them to work from home instead of commuting to work each day.

Consumer Products

Consumers are often poorly informed of the savings of energy efficient products. A prominent example of this is the energy savings that can be made by replacing an incandescent light bulb with a more modern alternative. When purchasing light bulbs, many consumers opt for cheap incandescent bulbs, failing to take into account their higher energy costs and lower lifespans when compared to modern compact fluorescent and LED bulbs. Although these energy-efficient alternatives have a higher upfront cost, their long lifespan and low energy use can save consumers a considerable amount of money. The price of LED bulbs has also been steadily decreasing in the past five years due to improvements in semiconductor technology. Many LED bulbs on the market qualify for utility rebates that further reduce the price of purchase to the consumer. Estimates by the U.S. Department of Energy state that widespread adoption of LED lighting over the next 20 years could result in about \$265 billion worth of savings in United States energy costs.

An assortment of energy-efficient semiconductor (LED) lamps for commercial and residential lighting use. LED lamps use at least 75% less energy, and last 25 times longer, than traditional incandescent light bulbs.

The research one must put into conserving energy is often too time consuming and costly for the average consumer when there are cheaper products and technology available using today's fossil fuels. Some governments and NGOs are attempting to reduce this complexity with ecolabels that make differences in energy efficiency easy to research while shopping.

To provide the kind of information and support people need to invest money, time and effort in energy conservation, it is important to understand and link to people's topical concerns. For instance, some retailers argue that bright lighting stimulates purchasing. However, health studies have demonstrated that headache, stress, blood pressure, fatigue and worker error all generally increase with the common over-illumination present in many workplace and retail settings. It has been shown that natural daylighting increases productivity levels of workers, while reducing energy consumption.

In warm climates where air conditioning is used, any household device that gives off heat will result in a larger load on the cooling system. Items such as stoves, dish washers, clothes dryers, hot water and incandescent lighting all add heat to the home. Low-power or insulated versions of these devices give off less heat for the air conditioning to remove. The air conditioning system can also improve in efficiency by using a heat sink that is cooler than the standard air heat exchanger, such as geothermal or water.

In cold climates, heating air and water is a major demand on household energy use. Significant energy reductions are possible by using different technologies. Heat pumps are a more efficient alternative to electrical resistance heaters for warming air or water. A variety of efficient clothes dryers are available, and the clothes lines requires no energy- only time. Natural-gas (or biogas) condensing boilers and hot-air furnaces increase efficiency over standard hot-flue models. Standard electric boilers can be made to run only at hours of the day when they are needed by means of a time switch. This decreases energy use vastly. In showers, a semi-closed-loop system could be used. New construction implementing heat exchangers can capture heat from waste water or exhaust air in bathrooms, laundry and kitchens.

In both warm and cold climate extremes, airtight thermal insulated construction is the largest factor determining the efficiency of a home. Insulation is added to minimize the flow of heat to or from the home, but can be labor-intensive to retrofit to an existing home.

Energy Conservation around the World

Asia

Despite the vital role energy efficiency is expected to play in cost-effectively cutting energy demand, only a small part of its economic potential is exploited in the Asia. Governments have implemented a range of subsidies such as cash grants, cheap credit, tax exemptions, and co-financing with public-sector funds to encourage a range of energy-efficiency initiatives across several sectors. Governments in the Asia-Pacific region have implemented a range of information provision and labeling programs for buildings, appliances, and the transportation and industrial sectors. Information programs can simply provide data, such as fuel-economy labels, or actively seek to encourage behavioral changes, such as Japan's Cool Biz campaign that encourages setting air conditioners at 28-degrees Celsius and allowing employees to dress casually in the summer.

European Union

At the end of 2006, the European Union (EU) pledged to cut its annual consumption of primary energy by 20% by 2020. The 'European Union Energy Efficiency Action Plan' is long-awaited. Directive 2012/27/EU is on energy efficiency.

As part of the EU's SAVE Programme, aimed at promoting energy efficiency and encouraging energy-saving behavior, the Boiler Efficiency Directive specifies minimum levels of efficiency for boilers utilizing liquid or gaseous fuels.

India

The Petroleum Conservation Research Association (PCRA) is an Indian governmental body created in 1978 that engages in promoting energy efficiency and conservation in every walk of life. In the recent past PCRA has done mass-media campaigns in television, radio, and print media. This is an impact-assessment survey by a third party revealed that due to these larger campaigns by PCRA, the public's overall awareness level has gone up leading to saving of fossil fuels worth crores of rupees, besides reducing pollution.

The Bureau of Energy Efficiency is an Indian government organization created in 2001 that is responsible for promoting energy efficiency and conservation.

Protection and Conservation of Natural Resources is done by CNRM Community Natural Resources Management.

Japan

Since the 1973 oil crisis, energy conservation has been an issue in Japan. All oil-based fuel is imported, so domestic sustainable energy is being developed.

The Energy Conservation Center promotes energy efficiency in every aspect of Japan. Public entities are implementing the efficient use of energy for industries and research. It includes projects such as the Top Runner Program. In this project, new appliances are regularly tested on efficiency, and the most efficient ones are made the standard.

Advertising with high energy in Shinjuku.

Lebanon

In Lebanon and since 2002 The Lebanese Center for Energy Conservation (LCEC) has been promoting the development of efficient and rational uses of energy and the use of renewable energy at the consumer level. It was created as a project financed by the International Environment Facility (GEF) and the Ministry of Energy Water (MEW) under the management of the United Nations Development Programme (UNDP) and gradually established itself as an independent technical national center although it continues to be supported by the United Nations Development Programme (UNDP) as indicated in the Memorandum of Understanding (MoU) signed between MEW and UNDP on 18 June 2007.

Nepal

Until recently, Nepal has been focusing on the exploitation of its huge water resources to produce hydro power. Demand side management and energy conservation was not in the focus of government action. In 2009, bilateral Development Cooperation between Nepal and the Federal Republic of Germany, has agreed upon the joint implementation of "Nepal Energy Efficiency Programme". The lead executing agencies for the implementation are the Water and Energy Commission Secretariat (WECS). The aim of the programme is the promotion of energy efficiency in policy making, in rural and urban households as well as in the industry. Due to the lack of a government organization that promotes energy efficiency in the country, the Federation of Nepalese Chambers of Commerce and Industry (FNCCI) has established the Energy Efficiency Centre under his roof to promote energy conservation in the private sector. The Energy Efficiency Centre is a non-profit initiative that is offering energy auditing services to the industries. The Centre is also supported by Nepal Energy Efficiency Programme of Deutsche Gesellschaft für Internationale Zusammenarbeit. A study conducted in 2012 found out that Nepalese industries could save 160,000 Megawatt hours of electricity and 8,000 Terajoule of thermal energy (like diesel, furnace oil and coal) every year. These savings are equivalent to annual energy cost cut of up to 6.4 Billion Nepalese Rupees. As a result of Nepal Economic Forum 2014, an economic reform agenda in the priority sectors was declared focusing on energy conservation among others. In the energy reform agenda the government of Nepal gave the commitment to introduce incentive packages in the budget of the fiscal year 2015/16 for industries that practices energy efficiency or use efficient technologies (incl. cogeneration).

New Zealand

In New Zealand the Energy Efficiency and Conservation Authority is the Government Agency responsible for promoting energy efficiency and conservation. The *Energy Management Association of New Zealand* is a membership based organization representing the New Zealand energy services sector, providing training and accreditation services with the aim of ensuring energy management services are credible and dependable.

Nigeria

In Nigeria, the Lagos State Government is encouraging Lagosians to imbibe an energy conservation culture. The Lagos State Electricity Board (LSEB) is spearheading an initiative tagged "Conserve Energy, Save Money" under the Ministry of Energy and Mineral Resources. The initiative is designed to sensitize Lagosians around the theme of energy conservation by connecting with and influencing their behavior through do-it-yourself tips and exciting interaction with prominent personalities. In September 2013, Governor Babatunde Raji Fashola of Lagos State and rapper Jude 'MI' Abaga (campaign ambassador) participated in the Governor's first ever Google+ Hangout on the topic of energy conservation.

In addition to the hangout, during the month of October (the official energy conservation month in the state), LSEB hosted experience centers in malls around Lagos State where members of the public were encouraged to calculate their current household energy consumption and discover ways to save money using the 1st-ever consumer-focused energy app in sub-saharan Africa. To get Lagosians started on energy conservation, Solar Lamps and Phillips Energy-saving bulbs were also given out at each experience center.

In Kaduna State, the Kaduna Power Supply Company (KAPSCO) ran a program to replace all light bulbs in Public Offices; fitting energy-saving bulbs in place of incandescent bulbs. KAPSCO is also embarking on an initiative to retrofit all conventional streetlights in the Kaduna Metropolis to LEDs which consume much less energy.

Sri Lanka

Sri Lanka currently consumes fossil fuels, hydro power, wind power, solar power and dendro power for their day to day power generation. The Sri Lanka Sustainable Energy Authority is playing a major role regarding energy management and energy conservation. Today, most of the industries are requested to reduce their energy consumption by using renewable energy sources and optimizing their energy usage.

Turkey

Turkey aims to decrease by at least 20% the amount of energy consumed per GDP of Turkey by the year 2023 (energy intensity).

United States

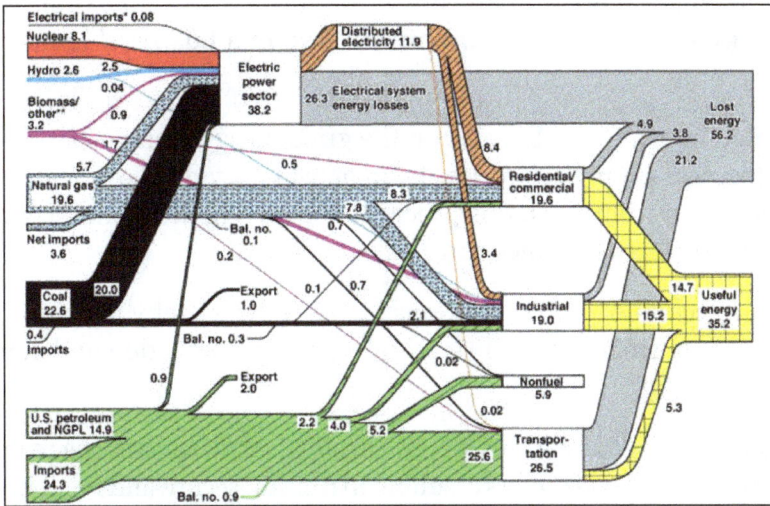

The United States is currently the second largest single consumer of energy, following China. The U.S. Department of Energy categorizes national energy use in four broad sectors: transportation, residential, commercial, and industrial.

Energy usage in transportation and residential sectors, about half of U.S. energy consumption, is largely controlled by individual consumers. Commercial and industrial energy expenditures are determined by businesses entities and other facility managers. National energy policy has a significant effect on energy usage across all four sectors.

Another aspect of energy conversation is using Leadership in Energy and Environmental Design. (LEED) This program is not mandatory, it is voluntary. This program has many categories, Energy and Atmosphere Prerequisite, applies to energy conservation. This topic focuses on energy performance, renewable energy, energy performance, and many more. This program is designed to promote energy efficiency and be a green building, which is part of conservation.

Energy Storage

Energy storage, sometimes referred to as electricity storage, is as the name suggests, the storage of energy. It works by capturing electricity produced by both renewable and non-renewable resources and storing it for discharge when required. The solution allows users to come off the grid and switch to stored electricity, at a time most beneficial, giving greater flexibility and control of electrical usage.

The electrical energy grid requires a balance between supply and demand. At times of low demand, when there is excess supply, it can be stored for use at times of high demand, with low supply, thus adjusting to provide the required balance between supply and demand.

This approach is especially effective with renewable generation, which is intermittent by its nature. Solar and wind, for example, generate little amounts of power in the absence of sunshine or wind. Energy storage is able to smooth out the supply from these sources to provide a more reliable supply that matches demand. This allows organisations to maximise renewable generation, therefore energy storage can play an integral role in a business' journey towards carbon neutrality.

At times of unexpected increases in demand on the grid, energy storage can be used to discharge power back to the electrical supply network very quickly to provide additional supply to help meet demand. By businesses contributing to this process of balancing the demand it alleviates the pressure from the grid and for this assistance contracts are offered.

Driven by advances in technology, the traditional model of electricity provision is being replaced by a smart, flexible energy smart grid powered by energy storage, demand side response (DSR) and inter-connectivity.

This responsive system will provide a balance between supply and demand via responsive power generation from both suppliers and consumers resulting in a clean, secure and reliable electricity supply.

Benefits of Energy Storage

Energy storage is key to achieving overarching low carbon and electrical network efficiency targets by:

- Deferring or avoiding investment in network reinforcement.
- Reducing the need for conventional generation, including peaking power plants.
- Meeting binding targets with lower renewable capacity.
- Maximising the use of low carbon, inflexible generation.
- Optimising balancing of the system on a minute by minute basis.

At grid level, energy storage reduces stress on the electrical network infrastructure, increases the proportion of renewables on the grid and increases reliability of renewable generation. It also provides efficient demand balancing options for the grid and reduces the need for backup demand generation.

For large electricity consumers, energy storage provides flexibility in electricity supply and opportunities for significant costs savings by enabling a switch to stored electricity at peak-tariff periods. It eliminates the risk of network interruption by providing full UPS capabilities, reducing the likelihood of energy related failures which can total as much as 17% of annual revenues and maximises the investment into renewable generation.

Energy Storage Technologies

Different energy storage technologies contribute to electricity stability by working at various stages of the grid, from generation to consumer end-use.

Thermal Storage

Thermal storage is used for electricity generation by using power from the sun, even when the sun is not shining. Concentrating solar plants can capture heat from the sun and store the energy in water, molten salts, or other fluids. This stored energy is later used to generate electricity, enabling the use of solar energy even after sunset.

Plants like these are currently operating or proposed in California, Arizona, and Nevada. For example, the proposed Rice Solar Energy Project in Blythe, California will use a molten salt storage system with a concentrating solar tower to provide power for approximately 68,000 homes each year.

Concentrating solar plants focus the sun's heat to store energy in water, molten salts, or other fluids, which can be utilized even after the sun has set.

Thermal storage technologies also exist for end-use energy storage. One method is freezing water at night using off-peak electricity, then releasing the stored cold energy from the ice to help with air conditioning during the day.

For example, Ice Energy's Ice Bear system creates a block of ice at night, and then uses the ice during the day to condense the air conditioning system's refrigerant. In this way, the Ice Bear system shifts the building's electricity consumption from the daytime peak to off-peak times when the electricity is less expensive. Additionally, the Bonneville Power Administration is conducting a pilot program on storing excess wind generation in residential water heaters.

Compressed Air

Compressed Air Energy Storage (CAES) also works as a generation storage technology by using the elastic potential energy of compressed air to improve the efficiencies of conventional gas turbines.

CAES systems compress air using electricity during off-peak times, and then store the air in underground caverns. During times of peak demand, the air is drawn from storage and fired with natural gas in a combustion turbine to generate electricity. This method uses only a third of the natural gas used in conventional methods. Because CAES plants require some sort of underground reservoir, they are limited by their locations. Two commercial CAES plants currently operate in Huntorf, Germany and MacIntosh, Alabama, though plants have been proposed in other parts of the United States.

Hydrogen

Hydrogen can be used as a zero-carbon fuel for generation. Excess electricity can be used to create hydrogen, which can be stored and used later in fuel cells, engines, or gas turbines to generate electricity without producing harmful emissions. NREL has studied the potential for creating hydrogen from wind power and storing it in the wind turbine towers for electricity generation when the wind isn't blowing.

The Wind to Hydrogen Project at NREL studies the storage of wind energy as hydrogen.

Pumped Hydroelectric Storage

Pumped hydroelectric storage offers a way to store energy at the grid's transmission stage, by storing excess generation for later use.

Many hydroelectric power plants include two reservoirs at different elevations. These plants store energy by pumping water into the upper reservoir when supply exceeds demand. When demand exceeds supply, the water is released into the lower reservoir by running downhill through turbines to generate electricity.

With more than 22 GW of installed capacity in the United States, pumped hydro storage is the largest storage system operating today. However, the long permitting process and high cost of pumped storage makes further projects unlikely.

Flywheels

Flywheels can provide a variety of benefits to the grid at either the transmission or distribution level, by storing electricity in the form of a spinning mass.

The device is shaped liked a cylinder and contains a large rotor inside a vacuum. When the flywheel draws power from the grid, the rotor accelerates to very high speeds, storing the electricity as rotational energy. To discharge the stored energy, the rotor switches to generation mode, slows down, and runs on inertial energy, thus returning electricity to the grid.

A flywheel rotor, is spun at high speed to store electricity as rotational energy.

Flywheels typically have long lifetimes and require little maintenance. The devices also have high efficiencies and rapid response times. Because they can be placed almost anywhere, flywheels can be located close to the consumers and store electricity for distribution.

While a single flywheel device has a typical capacity on the order of kilowatts, many flywheels can be connected in a "flywheel farm" to create a storage facility on the order of megawatts. Beacon Power's Stephentown Flywheel Energy Storage Plant in New York is the largest flywheel facility in the United States, with an operating capacity of 20 MW.

Batteries

Batteries, like those in a flashlight or cell phone, can also be used to store energy on a large scale.

Like flywheels, batteries can be located anywhere so they are often seen as storage for distribution, when a battery facility is located near consumers to provide power stability; or end-use, like batteries in electric vehicles.

Batteries can be located in communities to provide power stability for homes.

There are many different types of batteries that have large-scale energy storage potential, including sodium-sulfur, metal air, lithium ion, and lead-acid batteries. There are several battery installations at wind farms; including the Notrees Wind Storage Demonstration Project in Texas, which uses a 36 MW battery facility to help ensure stability of the power supply even when the wind isn't blowing.

Advancements in battery technologies have been made largely due to the expanding electric vehicle (EV) industry. As more developments are made with EVs, battery cost should continue to decline. Electric vehicles could also have an impact on energy storage through vehicle-to-grid technologies, in which their batteries can be connected to the grid and discharge power for others to use.

Future of Energy Storage

As new energy storage technologies are researched and tested, some barriers are likely to slow the commercialization of these technologies.

Energy storage is expensive, especially without policies that place a monetary value on the unique benefits of storage. Plus there is no current need for additional storage capacity to maintain electricity grid reliability. Without an operational need, it is difficult for storage to be cost-effective in the present. Furthermore, storage lacks a robust track record of large commercial-scale projects (with the exception of pumped hydro), making it difficult to deploy new projects.

Despite these potential barriers, certain programs and policies can help drive the development and deployment of storage technologies. The Department of Energy's Energy Storage Program researches different storage technologies and works closely with industry on pilot storage programs.

Research programs can help advance the deployment and commercialization of energy storage.

The deployment of storage technologies can also be advanced through renewable electricity standards (RES). Some states recognize storage technologies as acceptable renewable generation in their RES, and other states award Renewable Energy Credits (REC) to energy generation from storage devices that were charged by renewables.

The Federal Energy Regulatory Commission (FERC), the agency that regulates the electricity grid, has created a pricing structure that pays storage technologies and other fast-ramping resources a higher price for their services. This pricing structure, called Pay-for-Performance, recognizes the value of rapid response in providing stability to the grid. Pay-for-Performance has the potential to make storage technologies more cost-effective on a commercial-scale. An investment tax credit (ITC) would also help accelerate the deployment of storage technologies.

With the support of government and industry, energy storage technologies can continue to develop and expand, aid in the increasing deployment of variable renewable energy sources, and help store an ever-growing amount of clean, renewable energy in the future.

References

- Energy-management-hierarchy: hierarchystructure.com, Retrieved 29 March, 2019

- R.Chattopadhyay (2014). Green Tribology, Green Surface Engineering and Global Warming. ASM International,USA. ISBN 1627080643

- Sustainableenergy: conserve-energy-future.com, Retrieved 16 January, 2019

- "Top 5 reasons to be energy efficient". Alliance to Save Energy (ASE). 20 July 2012. Retrieved 14 June 2016

- How-energy-storage-works, clean-energy: ucsusa.org, Retrieved 11 January, 2019

- What-is-energy-storage, virtue: powerstar.com, Retrieved 3 May, 2019

PERMISSIONS

We would like to thank the editorial team for lending their expertise to make the book truly unique. They have played a crucial role in the development of this book. Without their invaluable contributions this book wouldn't have been possible. They have made vital efforts to compile up to date information on the varied aspects of this subject to make this book a valuable addition to the collection of many professionals and students.

This book was conceptualized with the vision of imparting up-to-date and integrated information in this field. To ensure the same, a matchless editorial board was set up. Every individual on the board went through rigorous rounds of assessment to prove their worth. After which they invested a large part of their time researching and compiling the most relevant data for our readers.

The editorial board has been involved in producing this book since its inception. They have spent rigorous hours researching and exploring the diverse topics which have resulted in the successful publishing of this book. They have passed on their knowledge of decades through this book. To expedite this challenging task, the publisher supported the team at every step. A small team of assistant editors was also appointed to further simplify the editing procedure and attain best results for the readers.

Apart from the editorial board, the designing team has also invested a significant amount of their time in understanding the subject and creating the most relevant covers. They scrutinized every image to scout for the most suitable representation of the subject and create an appropriate cover for the book.

The publishing team has been an ardent support to the editorial, designing and production team. Their endless efforts to recruit the best for this project, has resulted in the accomplishment of this book. They are a veteran in the field of academics and their pool of knowledge is as vast as their experience in printing. Their expertise and guidance has proved useful at every step. Their uncompromising quality standards have made this book an exceptional effort. Their encouragement from time to time has been an inspiration for everyone.

The publisher and the editorial board hope that this book will prove to be a valuable piece of knowledge for students, practitioners and scholars across the globe.

INDEX